Basic Hydraulics

Basic Hydraulics

Andrew L. Simon
The University of Akron

JOHN WILEY & SONS

New York
Chichester
Brisbane
Toronto

Library of Congress Cataloging in Publication Data:

Simon, Andrew L
  Basic hydraulics.

  Bibliography: p.
  Includes index.
    1. Hydraulics. I. Title.
TC160.S378   627'.042   80-15341
ISBN 0-471-07965-0

Printed in the United States of America

10 9 8 7 6 5 4 3 2 1

In remembrance of my mother

Preface

This book attempts to fill a need for an introductory hydraulics text broad enough to provide adequate coverage for diverse curricular needs. More material is included than is necessary for a three-semester credit course and it was planned this way. Some instructors will probably select Chapters 2, 3, and 4 on the properties of water and the basics on hydrostatics and hydrodynamics, followed by Chapters 6 and 7 on pipe flow and pumps, then conclude the course by selecting some parts of other chapters. Others will put more emphasis on Chapters 1, 8, 9, and 10, in addition to the basic topics, to orient their program toward natural watercourses. A serious attempt was made when writing each chapter to make it independent of the others.

The main feature of this book is its use of the metric S.I. system, although English equivalents are also shown. It was found to be much easier to teach the fluid mechanics fundamentals without the intellectual quagmire of ''pound–mass, pound force, slugs,'' and so on, and once the basics are taken care of, the continued use of metrics follows naturally. Many well-known nomographs and charts were converted to metrics and new ones were included from the most recent European works. For those wishing to work in the traditional English system, Appendix 1 includes adequate conversion data.

The author wishes to express his appreciation to the many users of his *Practical Hydraulics* who came forward with helpful recommendations and criticisms. This book was a direct result of some of their comments. Sincere gratitude is due to the reviewers, Mr. Dennis Carrigan, Texas State Technical Institute, Helm Haas, San Joaquin Delta College, and Dr. Vladimir Novotny, Marquette University, for their careful study of the manuscript. Special thanks are due to Mr. Richard Yen for working out most examples and problems. The reliable typing by Mrs. Agnes Stitz and Mrs. Dorothy Guilliams is deeply appreciated.

ANDREW L. SIMON

Contents

Basic Hydraulics

Chapter 1
Hydrology

1.1. Weather

Water is by far the most abundant substance on the earth's surface. Oceans cover over 70 percent of the earth. We live on the bottom of a sea of air, the atmosphere. The lower three-fourths of the atmosphere is dense enough to support some moisture. This layer is called *troposphere,* and it is in constant motion. The everchanging state of this layer is what we call *weather.* The sun is the single noteworthy source of energy that causes the constant changes in the weather. Consequently, the equatorial regions of the earth receive the greatest amount of solar radiation, the air near the surface warms up, and the water of the ocean's surface evaporates. The warm moist air rises and colder air rushes in to replace it. The latitudinal difference in earth's heating produces differences in air pressure all over the surface resulting in a general pattern of air movement. This pattern is influenced by many forces: the rotation of the earth, geography, seasonal variations, and so on. The air tends to equalize the pressures caused by uneven heating of the surface. From a cold high pressure region the winds blow toward a warm and moist low pressure region.

In areas of high pressure when air remains for a significant amount of time over a fairly uniform surface, the air takes on the characteristics of the surface: the coldness and dryness of polar lands or the humidity and warmth of the tropics. The air retains these characteristics for a long time, even after it has been moved to other regions. Such air masses bring about major changes in the weather, wherever they are carried by the winds.

The temperature of the troposphere decreases with height. As the moist air rises it tends to reach the dew point causing the formation of clouds. Typically, warm air masses contain a lot of moisture; hence they are characterized by high humidity and lots of clouds. Dry cold air masses are cloudless. The density of air decreases with the increase of temperature. Therefore, cold air masses represent high pressure while warm air masses result in low pressure at the earth's surface. The daily weather report shows the location of these *highs* and *lows*. Where these air masses meet there is a *weather front*. Where the high pressure cold air mass replaces the low pressure warm air mass, there is a *cold front*. The heavy cold air intrudes under the warm air and displaces it. In raising the warm moist air the water condenses and precipitates from the clouds creating rain or snow. Figure 1.1*a* is a schematic drawing of a cold front. Cold fronts move relatively fast. The angle of rise of the warm air in front of them is steep; therefore the rains are short, intensive, and relatively isolated. When the warm air mass displaces the cold air, there is a *warm front*. Warm fronts have a rather low angle of rise when compared to cold fronts. There the warm air shifts over the cold air squeezing it out. The warm front moves much slower than the cold front. Its geographic extent is wide. As a result rains caused by the movement of a warm front are extensive, low in intensity, and may last for days. Over the continental United States weather is generally influenced by cold air masses moving over the

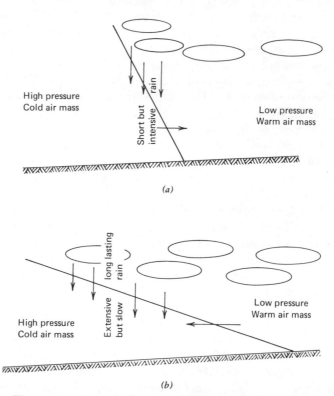

Figure 1.1   Weather fronts: (a) cold front; (b) warm front.

Great Plains from the northwest and by moist warm air masses moving north from the Gulf of Mexico.

Geographic conditions also contribute to the distribution of rainfall. The Rocky Mountains, for example, cause the rising of the clouds as they travel toward the east, which results in heavy rainfalls on the western slope and frequent lack of rains on the eastern slope. The warmth of the Great Lakes' waters during the winter months results in a rise of the northbound warm air masses, causing heavy local snowfalls near the shores. The combined influence and interdependence of many effects determine the quantity and the seasonal and geographic distribution of rainfall over much of the United States.

1.2. The Hydrologic Cycle

A portion of the water that is carried as atmospheric moisture falls on the land in the form of rain, snow, dew, or hail. Of these, rain is the most significant. Life on land would not be possible without it. As rain falls on the ground, part of it soon evaporates, rejoining the moisture of the atmosphere. Another part infiltrates into the ground. When in the ground it may be sucked up—transpired—by the vegetation, then evaporated through the leaves. Evapotranspiration by plants returns a fraction of the water in the ground to the atmosphere. The remainder of the ground water enters the open pores of the soil layers and begins its slow movement downward, back toward the sea. The third part of the precipitation runs off the ground, collects into rivulets, creeks, and rivers, and returns to the oceans. This circulation from oceans, atmosphere, ground, and back to the oceans is called the *hydrologic cycle*. Its essence is shown in the flow diagram in Figure 1.2. Of course this cyclic movement is not an orderly continuous process. Rains fall only intermittently; evaporation after a rain is a matter of a few hours. In colder climates snow that has fallen may not melt and run off for months. Some "innate waters" are held in underground layers indefinitely.

The quantitative evaluation of the flow in the hydrologic cycle for small regions or individual watersheds is an important aspect of hydraulic design. The amount of rainfall that may be reasonably expected, the percentage of the rainfall that will collect on the surface in the form of overland flow and will eventually show up as a flood wave in a river, the yearly amount of evaporation from a reservoir, the amount of rainfall that will recharge a much pumped ground water aquifer, are only some of the questions hydrologists are often called to answer.

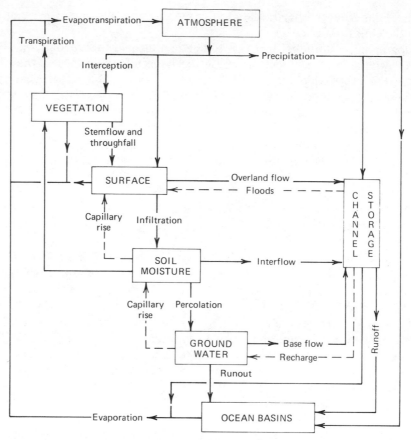

Figure 1.2 The hydrologic cycle.

Because of its random and intermittent nature, the movement of water through the paths of the hydrologic cycle can be studied by statistical means only. Before any reliable data are supplied for practical purposes, decades of careful collection of weather information, rainfall measurements, flood stage and discharge studies must be done. Even when such data are available, the results are somewhat illusory from the standpoint of the designer. Even when statistical theory is capable of predicting the expected magnitude of a "50 year flood" based on a mere 30 years of data, there is no assurance that a 100 year monster flood will not appear in the second year of a project's existence. For these reasons it is important that a hydraulician develop a basic understanding of the fundamentals of statistics used in hydrology.

1.3. Statistical Concepts

There are many apparent interdependences in hydrology that cannot be expressed as mathematical relationships. For instance, the total rainfall measured during a month in Pittsburgh, Pa., is clearly related to the average discharge of the Ohio River at Cincinnati, Ohio, some time later. But the discharge is a function of many other things, most of which are not as well defined. One could, however, plot such statistically interdependent pairs of variables in a coordinate system, as shown in Figure 1.3. If a sufficient number of variable-pairs are plotted, a certain trend may be observed. This trend may be strong enough to allow the drawing of a line of "best fit," which would represent a quasi-mathematical relationship between the two variables. The equation between two statistically interdependent variables is called a *regression equation*. If the scatter of the points around the line drawn is small, that is, the points are grouped close to the line of best fit, one speaks of a good *correlation*. When the scatter is wide, the correlation is poor. The degree of correlation is expressed by a coefficient ranging from zero to one. A zero correlation coefficient indicates that there is absolutely no interdependence between the two variables. If the correlation coefficient is unity, there is an apparently strict mathematical relationship between the two variables. In such a case there is no scatter of points about the line represent-

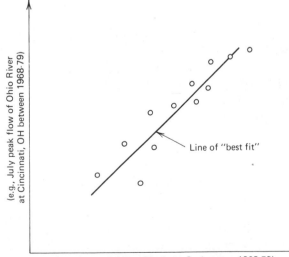

(e.g., June rainfall at Pittsburg, Pa. between 1968-79)

Figure 1.3 Regression analysis of statistically independent variables.

ing the relationship. There are "canned" computer programs available to perform such regression analysis on any type and number of data. If there are several interdependent variables, one must perform a *multiple regression analysis*. An example of such an analysis is the relationship between flood discharge, monthly rainfall, watershed area, channel slope, ground cover, and so on. The more pertinent variables one includes into the analysis, the closer the final results will fit, and the higher the correlation.

One of the most important problems facing hydrologists is the determination of the largest amount of rainfall that may be expected over a particular watershed. Rainfall is measured by standard rainfall gages at most airports and at many other locations throughout the world. The data available are often the amount of rainfall that has fallen during a 24 hour period unless continuous recorders are used. Data series extending over decades, and in some places well over a century, exist for many locations within the United States. The source of most of this data is the U.S. Weather Bureau. Much of the available data has been statistically analyzed. The results are often published in the form of *climatological atlases*. These usually show the maximum amount of daily rainfall that may be expected to occur once in 5, 10, 50, or 100 years, and so on. This type of information is based on the *theory of probability*.

The statistical theory of probability deals with the analysis of random variables. Let us consider the case of flipping coins. The best likelihood is that one gets a series of alternating *heads* and *tails*, but it is no way certain. The likelihood is quite small that one has a series of five *heads* before one's luck turns, and it is very remote, although possible, to have a run on *tails* in a continuous series of ten. The *frequency* of continuous runs of either heads or tails may be plotted against their number of occurrence in a graph like the one shown in Figure 1.4. The bell-shaped plot is known as the Gaussian probability curve. While both rainfalls and stream discharges can be treated as random variables, they cannot be analyzed by the Gaussian approach. One of the many reasons for this is that their probability distributions are skewed and bound at the low end; there is no such thing as negative stream flow or less than zero rainfall. One of the several probability distributions that was found to fit hydrological data is the double-logarithmic probability function proposed by Gumbel. It is based on the *theory of extreme values*. It relates the magnitude of the largest daily rainfall of a year within a series of years in which measurements are available to the frequency of its occurrence during the series. Likewise, it may relate the largest flood of each year to the frequency of occurrence during the period measured. For instance, the five largest of the yearly rainfalls during a measured period of 50 years have a frequency of 10 years, even if they all occurred during the last consecutive 5

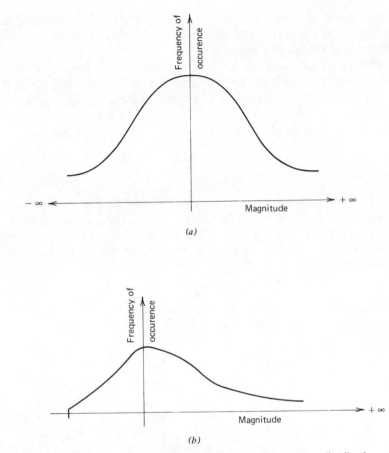

Figure 1.4 Frequency distributions: (a) Gaussian frequency distribution; (b) skewed frequency distribution bounded at the lower end.

years of the period. Generally, if the number of data during the period studied is n (50 in the previous case) and the rainfall has exceeded a certain Z value m times during this period, then the frequency of occurrence of Z rainfall is

$$p \, (frequency) = \frac{m}{n+1} \, 100 \qquad (1.1)$$

An alternate expression for the frequency is the *recurrence interval,*

$$T \, (years) = \frac{n+1}{m} \qquad (1.2)$$

which is the period of time, in years, within which the rainfall equal to or exceeding Z may be expected. The recurrence of floods of certain magnitude may be analyzed in the same manner.

The mathematical analysis required in the use of the theory of extreme values, or Gumbel's method, is far too complex to be included here. However, if one plots the functional values of Gumbel's probability equation just like a semilogarithmic graph, then the practical use of the method becomes a very simple matter. The following example will demonstrate the method.

Example 1.1

Using Gumbel's Theory of Extreme Values, determine the 5, 10, and 20 year rainfalls on the basis of an eleven year series of measurements of daily rainfalls. From the daily data obtained from a local airport yearly maximums were selected. The yearly maximums are shown below in the first two columns:

Year	Rainfall (mm)	Ranking order (m)	Return period (T) [Eq. (1.2)]	Frequency (p) [Eq. (1.1)]
1969	37	7	1.72	58
1970	20	11	1.09	95
1971	32	8	1.5	66.6
1972	60	3	4.0	25
1973	25	9	1.33	75.2
1974	52	4	3.0	33.3
1975	46	6	2.0	50
1976	70	2	6.0	16.6
1977	92	1	12.0	8.3
1978	48	5	2.4	41.6
1979	24	10	1.4	71.4

Solution

After the maximum yearly rainfalls are listed, they should be ranked in decreasing order of magnitude. The ranking order is shown in the third column of the table. It indicates that the largest rainfall of the period occurred during 1977. The number of years in the period is $n = 11$. Using Equation 1.2 the return periods may be computed as shown in column four. Column five lists the frequency as computed by Equation 1.1.

Assuming that Gumbel's theory holds, the data will plot as a straight line on a Gumbel graph. Figure E1.1 shows such a graph. The data from columns two and five (or four) are shown plotted. The line of best fit is drawn through the data points. To obtain the 5, 10, and 20 year rainfalls, start from the top of the chart and turn at the line to read off the corresponding rainfalls at the left. They are 69, 86, and 100 millimeters (2.7, 3.4, and 3.9 in.), respectively.

The advantage in using probability charts like the one shown in Figure E1.1 is that once the data are plotted on it, it could be extended by linear extrapolation to provide design values for return periods far exceeding the data range. While theoretically possible and frequently used, such extrapolation may lead to significant errors. The shorter the basic data period, the greater the potential error. Generally, data periods of less than 10 years are of little statistical value. When longer data ranges are not available at the site, the data should be supplemented by considering several other sets collected at other locations in the vicinity. The compounding of such data is beyond the scope of this book.

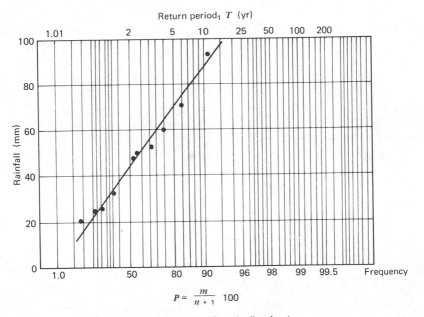

$$P = \frac{m}{n+1} \, 100$$

Figure E1.1 Gumbel's chart.

1.4. Rainfall

Rainfalls are defined by three parameters: duration, intensity, and frequency. The *duration* is the period of time during which a particular rain is falling. It may be expressed in minutes, hours, or days. The amount of water that has fallen during a rain is measured by rainfall gages and is expressed in *rainfall depths*. In the case of snowfall the water is melted so that it can be expressed in terms of equivalent rainfall. Dividing the rainfall depth with the duration of the rain results in *rainfall intensity*. It is expressed in millimeters per hour (or inches per hour). Because of the nature of rains, rainfalls of short duration usually are of high intensity while extended rains usually are of low intensity. This comes from the nature of cold and warm fronts, described earlier. Frequency of a rainfall was discussed in the previous section, but that discussion was limited to rains measured during 24 hour periods. Statistical analysis could be performed for rains of various durations. Figure 1.5 shows the results of one such analysis where the duration, frequency, and rain fall depth are related. These data are based on measurements at the Akron–Canton Airport in Ohio.

The purpose of the analysis of rainfalls is usually to determine the magnitude of runoff from a given watershed. The rainfall data provided in the form of intensity, frequency, and duration graphs usually refer to rains measured at a gaging station. It is called *point rainfall*. Because of the nature of rains, it is obvious that data on point rainfalls cannot be extended indiscriminately over a large watershed. This is particularly so with rains of high intensity that are usually of short duration and of very limited extent. To convert point rainfalls to probable *area rainfalls,* the graph shown in Figure 1.6 may be used. This chart was prepared on the basis of rainfall statistics gathered from all over the continental United States. It also indicates that rainfalls of long duration can be expected to cover much larger areas than those of short duration.

Example 1.2

Determine the intensity of a rainfall expected near Akron, Ohio, corresponding to a half hour rain with a return period of 5 years.

Solution

Using the corresponding rainfall statistics shown in Figure 1.5, enter the chart at a 30 minute duration. Turning at the 5 year frequency line the intensity of the rain is found to be 65 mm/hr (2.6 in./hr). In terms of rainfall depth this represents 32.5 mm (1.3 in.) of rain.

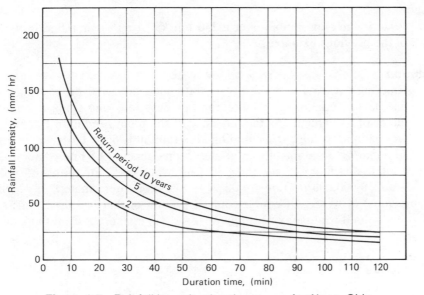

Figure 1.5 Rainfall intensity-duration curves for Akron, Ohio.

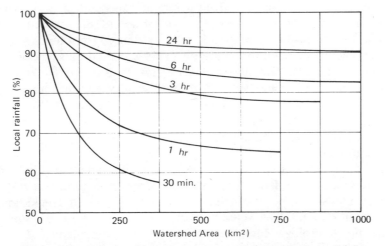

Figure 1.6 Relationship between area rainfall and point rainfall.

Example 1.3

What is the area rainfall over a 250 km² (97.7 mi²) watershed in the case of the previous example?

Solution

Entering Figure 1.6 at the given watershed area and proceeding to the line representing a 30 minute rainfall duration, one finds a reduction factor of 61 percent. As the intensity of the point rainfall in Example 1.2 was 65 mm/hr, the corresponding area rainfall is 65 × 0.61, which equals 39.6 mm/hr. In terms of rainfall depth this means 19.8 mm or about 2 cm rainfall distributed over the 250 km² watershed. In terms of volume it is 250,000,000 m² times 0.02 m = 5 million cubic meters (1.32 billion gallons) of water. Of course, only a portion of this will be runoff; some of it will infiltrate into the soil and some will evaporate from the land.

1.5 Floods

Rainfall is only one of several parameters that contribute to the magnitude of floods encountered at a particular site. Other important parameters include the size of the land area from where the rainfall is collected, the shape of this area, and its average slope along the main channel through which the rainwater is led to the site in consideration. The type of soil (permeable or impermeable), cultivation, land use, and similar factors all contribute to the relative magnitude of a flood. Of all these variables the most important one for flood magnitude is the land area from where all rains collect to the site of the structure designed. This is called the *watershed* of the particular outflow point. The size of this drainage area is to be measured from topographic maps by the designer. Distinction is made by experts of hydrology between small and large watersheds. Drainage areas of less than about 80 km² (31 mi²) are considered small watersheds. For small watersheds the most important factor after the size of the drainage area is the rainfall intensity. Second to this is the slope of the main channel. For large watersheds the most important variable after the drainage area size is the main channel slope, followed by the rainfall intensity.

Once the rain has started to fall, it will begin its pattern of flow along the steepest descent of the land. The rate of collection depends on the relative density of available flow channels. Water flows faster in creeks and other channels than on land. The velocity of the flow depends on the slope of the channels. For hydrologic computations the term "main channel slope," S, is the ratio of the difference in elevation between the points 10 and 85 percent of the total drop of elevation upstream from the studied site to the watershed boundary divided by the length of channel between these points. Hence the steepest 15 percent and the flattest 10 percent of the land contours are excluded in this consideration. The main channel slope is determined by the design on the basis of available topographic maps.

The time it takes for the first raindrop that falls at the most distant point of the drainage area to reach the outlet of the watershed is called the *time of concentration, T_c.* Experience has shown that the most critical floods are of rains whose duration at least equals the time of concentration. For small drainage basins of agricultural land the time of concentration in minutes may be estimated from the formula

$$T_c = 0.022 K^{0.77} = 0.022 \left[\frac{L}{\sqrt{S}} \right]^{0.77} \tag{1.3}$$

where K equals the maximum length of travel in meters divided by the square root of the main channel slope S.

The effect of surface soils and geologic characteristics is known to be highly important to flood peaks. Depending on the permeability of the land, part of the rainfall during its initial period infiltrates into the ground. Rainfall up to 12 mm (0.5 in.) may be absorbed into the soil in humid regions, provided the soil was relatively dry prior to the beginning of the rainfall. The rate of infiltration depends on the soil cover, land use, and cultivation. Soil infiltration rates are often available from the Soil Conservation Service, U.S. Department of Agriculture, and similar organizations.

The magnitude of floods may be estimated by what is known as the *rational formula,* Equation 1.4. It assumes a direct relationship between discharge, critical rainfall intensity, and drainage area, in the form of

$$Q_T = cI_TA \tag{1.4}$$

The critical rainfall intensity refers to a rain whose duration equals the time of concentration T_c representative of the watershed. The value of T_c may be approximated by Equation 1.3. The critical rainfall intensity then may be obtained from results of statistical rainfall analysis like the one shown in Figure 1.5. If the rainfall intensity is substituted into Equation 1.4 in millimeters per hour and the watershed area is measured in square kilometers, then the peak flood discharge resulting from Equation 1.4 will be in 1000 m³/hr. The c factor of proportionality is the runoff coefficient, a dimensionless quantity indicating the amount of surface runoff from different soil surfaces. Its values are listed in Table 1.1. The range of c values shown in the table indicates the great degree of uncertainty associated with the rational formula. But because of its simplicity the method is quite useful in designing small hydraulic installations like culverts, drainage structures, and the like. More complex methods for determining flood peaks are available, but they are beyond the scope of this text.

Table 1.1
Runoff Coefficients Used with the Rational Formula [Eq. (1.4)]

Description of Area	Runoff coefficients
Business	
Downtown	0.70 to 0.95
Neighborhood	0.50 to 0.70
Residential	
Single-family	0.30 to 0.50
Multi-units, detached	0.40 to 0.60
Multi-units, attached	0.60 to 0.75
Residential (surburban)	0.25 to 0.40
Apartments	0.50 to 0.70
Industrial	
Light	0.50 to 0.80
Heavy	0.60 to 0.90
Parks, cemeteries	0.10 to 0.25
Playgrounds	0.20 to 0.35
Railroad yard	0.20 to 0.35
Unimproved	0.10 to 0.30

Example 1.4

A 20 km² (7.8 mi²) watershed in the Akron, Ohio area is shaped such that the distance of the outlet point to the farthermost edge of the catchment area is 5000 m (5470 yd). The land is agricultural and the runoff coefficient is estimated to be 0.30. The main channel slope is 0.005. Determine the expected peak of the 5 year flood.

Solution

To use the rational method, the time of concentration T_c must first be determined by the use of Equation 1.3. In that equation the value of K is

$$K = \frac{5000}{(0.005)^{1/2}} = 7042$$

and $K^{0.77} = 917$. Hence Equation 1.3 will result in

$$T_c = 0.022K^{0.77} = 18 \text{ min}$$

The time of concentration will equal the duration of the critical rainfall for the flood frequency sought. Using Figure 1.5 the intensity of an 18 minute rain with 5 year frequency is

$$I_{18} = 87 \text{ mm/hr } (3.4 \text{ in./hr})$$

Since this is point rainfall, one may utilize Figure 1.6 for conversion. With some approximation a reduction factor of 0.85 may be taken, and the area rainfall intensity of

$$I = 0.85(87) = 74 \text{ mm/hr}$$

may be taken as critical rainfall. Substituting all known values into the rational formula, Equation 1.4, one obtains

$$Q_{5yr} = 0.30(74 \text{ mm/hr}) \, (20 \text{ km}^2) = 0.30(0.074) \, (20{,}000{,}000)$$

$$= 444{,}000 \text{ m}^3/\text{hr}$$

$$Q_{5yr} = 123 \text{ m}^3/\text{hr } (4367 \text{ cfs})$$

Problems

1.1 The yearly maximum floods of a creek during a 15 year period were measured as follows:

1964	68.0 m³/s	1972	61.3 m³/s
1965	50.5	1973	54.7
1966	82.5	1974	39.7
1967	102.4	1975	74.6
1968	42.3	1976	57.9
1969	55.7	1977	64.6
1970	76.6	1978	89.9
1971	34.7		

Determine the return period of the 1967 flood. (*Ans.* 16 years)

1.2 Using the data base of Problem 1.1, determine the 5, 10, and 25 year floods. (*Ans.* 80, 95, and 114 m³/s)

1.3 Assume that the 1979 maximum flood in the stream described in Problem 1.1 was 250 m³/s. In what sense would this change the return periods determined in Problem 1.2?

1.4 The rainfall depth measured during a 2 and one half hour rain was 84 mm. Compute the intensity of the rainfall. (*Ans.* 33.6 mm/hr)

1.5 Determine the rainfall depth of a 2 hour rain at Akron, Ohio, if the frequency of that rain is 10 years. (*Ans.* 50 mm)

1.6 If the watershed area is 125 km², what would be the area rainfall for the case described in Problem 1.5? (*Ans.* 40 mm)

1.7 The maximum length of travel in a watershed is 4700 m. The mean channel slope is 0.007. Determine the time of concentration. (*Ans.* 91 min)

1.8 A suburban residential area is 5 km². The time of concentration is 20 min. Using Figure 1.5 determine the expected peak discharge for the 10 year rainfall. (*Ans.* 35 to 56 m³/s)

Chapter 2
Properties
of
Water

2.1 The Three Phases of Water

Water is a chemical compound composed of oxygen and hydrogen. In each water molecule there is one oxygen atom and two hydrogen atoms. Clusters of water molecules are more or less bonded together by their hydrogen atoms. This type of atomic bond is called *hydrogen bond*. The degree of hydrogen bonding—the amount of energy holding the molecules together—depends on the temperature and the pressure present. Both temperature and pressure are manifestations of energy.

Depending on its *internal energy content,* water appears in either liquid, solid, or gaseous forms. Snow and ice are solid forms of water; moisture, water vapor in air, is a gaseous form. The different forms of water are called its phases. Whether water is in its solid, liquid, or gaseous phase depends on the amount of energy held in its hydrogen bonds. In solid form all hydrogen atoms are bonded; in liquid form fewer hydrogens are bonded. There are no bonds in the gaseous phase. Water is a stable compound; the bonding between the hydrogen and oxygen atoms does not decompose until the temperature reaches thousands of degrees Celsius.

The various phases of water as they depend on the energy content are shown in a phase diagram in Figure 2.1. The two coordinates of the phase diagram represent the two forms of energy: pressure and temperature. The diagram contains three zones within which water is in solid, liquid, or gaseous condition. To pass from one phase to another, either the pressure or the heat energy must change. Therefore energy is either added or taken away from water in order to change its phase.

For practical engineering purposes the state of water under atmospheric pressure and at normal average temperature—condition is considered to be standard—is of particular importance. The pressure of the atmosphere at sea

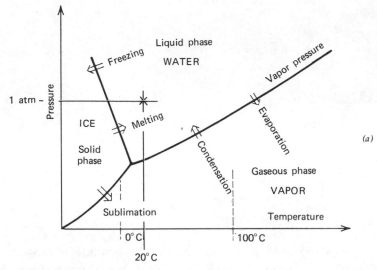

Figure 2.1 The three phases of water. The starred point represents the standard condition at 1 atm pressure and 20°C.

level at a latitude of 45 degrees and a temperature of 20 degrees Celsius is taken to be equal to one *atmosphere*. This equals the pressure exerted by a water column of 9.75 m (32 ft) height. For everyday computations this may be rounded up to 10 m (33 ft) of water. Therefore a 10-m high column of water defines 1 atm pressure. At higher elevations the atmospheric pressure is less. For every 713 m (2400 ft) of height above sea level, the atmospheric pressure is reduced by about 1 m (3.3 ft).

At normal atmospheric pressure water will freeze if its temperature is reduced to the zero degree and will turn into vapor (will boil) if heated to 100 degrees on the Celsius scale. At higher elevations where the atmospheric pressure is less, water will boil below 100°C. If the temperature of water is maintained at normal room temperature, at 20°C, and the pressure is reduced to the equivalent of 0.24 m (0.79 ft) of water, it will turn into vapor again. This is called *vapor pressure*. It is very important in the design of closed systems like pipes and pumps where water may change into vapor because of the reduction of pressure, even though the temperature remains constant. Failure to consider this aspect of water may lead to a damaging process called *cavitation*. This subject will be taken up in more detail in Chapter 7. Table 2.1 shows the vapor pressure of water with respect to temperature. Pressure values are shown in terms of Newtons/square meters (N/m²) as well as in terms of the height of a column of cold water.

Table 2.1
Vapor Pressure of Water

Temperature (°C)	Absolute pressure (N/m²)	Height of cold-water column (m)
0	613	0.06
10	1,220	0.12
20	2,330	0.24
40	7,340	0.74
60	19,900	2.04
80	47,400	4.87

Example 2.1

A pump is installed on a 3000-m (9840-ft) high mountain, where the atmospheric (barometric) pressure is 29 percent less than at sea level and the ambient pressure is 10°C. What will be the theoretical maximum height of a still water column in the suction pipe of the pump?

Solution

If at sea level the standard atmospheric pressure is 9.75 m of water, the atmospheric pressure at the pump's location will be 29 percent less, that is,

$$9.75 - 0.29(9.75) = 6.92 \text{ m}$$

From Table 2.1 the vapor pressure of water is 1226 N/m² or the pressure of a 0.12-m high column of water. To find the theoretical maximum height of the still water column in the pump's suction pipe we must reduce the ambient atmospheric pressure by the vapor pressure, that is,

$$6.92 - 0.12 = 6.8 \text{ m (22.3 ft)}$$

Note that the increase in elevation decreased the atmospheric pressure and that the decrease in temperature decreased the vapor pressure. If the pump in question would operate in hot water, the needed reduction due to vapor pressure would drastically reduce the theoretical maximum height of the water column in the suction pipe.

2.2 Density and Specific Weight

The mass of water in a unit volume is called its *density*. It is obtained by dividing the weight of water by its volume, that is, $\rho = W/V$. Its magnitude is dependent on the number of water molecules that occupy the space of a unit volume. This, of course, is determined by the size of the molecules and by the structure by which they bond together. The latter, as we know already, depends on the temperature and pressure. Because of the peculiar molecular structure of water and the change of the molecular structure when the water takes solid form, it is one of the few substances that expands when it freezes. The expansion of freezing water when rigidly contained causes stresses in the container. These stresses are responsible for the weathering of rocks and could damage pipes or structures if their effect is not considered in the design.

Water reaches its maximum density near the freezing point at 3.98°C. Table 2.2 gives the density of water at different temperatures.

As shown in Table 2.2 the density of ice is different from that of liquid water at the same temperature. This is why ice floats on water.

Since sea water contains salt, its density is greater than that of fresh water. The density of sea water is usually taken to be about 4 percent more than that of fresh water.

Table 2.2
Density and Specific Weight of Water

Temperature (°C)	Density, ρ (kg/m³)	Specific weight, γ (N/m³)
0 (ice)	917	8,996
0 (water)	999	9,800
3.98	1,000	9,810
10	999	9,800
25	997	9,780
100	958	9,397

Example 2.2

The temperature of 0.5 m³ (17.7 ft³) water is 10°C. What will be the change in volume if the water is heated to 25°C?

Solution

Heating will not change the total mass, hence we can write that

$$\text{Volume} \times \text{Density} = \text{Mass} = \text{Constant}$$

or

$$V_1 \times \rho_1 = V_2 \times \rho_2 = \text{const}$$

Using Table 1.2 and denoting the initial state with subscript 1, we have

$$V_1 = 0.5 \text{ m}^3$$

$$\rho_1 = 999 \text{ kg/m}^3$$

$$\rho_2 = 997 \text{ kg/m}^3$$

From here

$$V_2 = V_1 \frac{\rho_1}{\rho_2} = 0.5 \frac{999}{997} = 0.501 \text{ m}^3$$

That is, the volume will increase by one liter (61.2 in.³) due to the increase in temperature.

Density changes due to pressure are assumed to be zero for almost all hydraulic calculations. In other words, water is generally assumed to be incompressible, even though it is about 100 times more compressible than steel. However, in computations relating to shock waves in water, knowledge of the elastic properties of water is essential.

The ratio of the change of pressure to the corresponding change of volume is called the *bulk modulus of elasticity*. In solid mechanics the modulus of elasticity (Young's modulus) is defined as the ratio of linear stresses to linear strains, determined by tension tests. With fluids we speak of bulk modulus because it is determined by compression tests on volumes. An initial volume of V_0 will change by ΔV by a change in surface pressure Δp. The formula expressing this relationship is

$$\Delta p = - E \cdot \frac{\Delta V}{V_0} \tag{2.1}$$

The elasticity of water at normal temperatures is generally taken to be 2.138 gigaNewton/m² (310,000 psi).

Example 2.3

An 0.5 m³ (17.7 ft³) volume of water is under 9.81 kN/m² (206 lb/ft²) pressure initially. If the pressure is increased to 88 kN/m² (1848 lb/ft²) and the temperature remains constant, how much will be the reduction in volume?

Solution

Equation 2.1 is to be used with the following known variables:

$$\Delta p = 88 - 9.81 = 78.19 \text{ kN/m}^2$$

$$V_0 = 0.5 \text{ m}^3$$

$$E = 2.138 \times 10^6 \text{ kN/m}^2$$

Substituting, we have

$$78.19 = -2.138 \times 10^6 \frac{\Delta V}{0.5}$$

from which

$$-\Delta V = \frac{0.5(78.19)}{2.138(10^6)} = 18.3 \times 10^{-6} \text{ m}^3$$

which is about a 18 cm³ (1.1 in.³) reduction in volume.

Density depends on the number of molecules contained in a certain volume. As such, its magnitude is independent of the location where it is measured. By virtue of *Newton's Second Law of Motion,* the weight of a substance is the product of its mass and the gravitational acceleration. For a unit mass, or density, the unit weight, or *specific weight* is then defined as

$$\gamma = \rho \cdot g \tag{2.2}$$

The gravitational acceleration on the earth's surface averages between 9.78 and 9.82 m/s². For design usually $g = 9.81$ m/s² (32.2 ft/sec²) is used. For rough computations 10 m/s² is commonly allowed, as this would introduce an error of about 2% only. Table 2.2 lists the value of the specific weight of water for various temperatures. For normal temperatures, the specific weight is taken as $\gamma = 9.81$ kN/m³ (62.4 lb/ft³). The alternate expression of *unit weight* is often applied in practice.

2.3 Viscosity: the Resistance to Flow

Perhaps the most important physical property of water is its resistance to shear or angular deformation. The measure of resistance of a fluid to such relative motion is called *viscosity.* We define viscosity as a capacity of a fluid to convert kinetic energy—energy of motion—into heat energy. The energy converted into heat is considered lost because it can no longer contribute to further motion. It either results in warming up the fluid or is lost into the atmosphere by dissipation. The energy required to move a certain amount of water through a pipe, open channel, or hydraulic structure is determined by the amount of *viscous shear losses* to be encountered along the way. Therefore viscosity of the fluid inherently controls its movement. Viscosity is due to the cohesion between fluid particles and also to the interchange of molecules between the layers of different velocities. Mathematically the relationship between viscous shear stress and viscosity is expressed by Newton's law of viscosity. This is written as

$$\tau = \mu \, \frac{\Delta v}{\Delta y} \tag{2.3}$$

which is nothing else but an expression of proportionality between viscous shear resistance τ and the rate of change of velocity Δv in the direction perpendicular to the shear stress Δy, as shown in Figure 2.2. The mechanism illustrated is in some ways similar to the case of a deck of cards dragged along on a table. The relative velocity between adjacent cards is Δv, the thickness of a card is Δy. The factor of proportionality is called *absolute* (or dynamic) *viscosity*, μ, and has the dimension of force per area (stress) times the time interval considered. In science, viscosity is usually measured in the unit of *poise* (P). One poise equals one hundred centipoises (cP). One poise also equals 0.1 Newton · seconds/square meter (0.3 (lb-sec) / in.²), or

$$1 \text{ cP} = 0.001 \, \frac{N \cdot s}{m^2} \tag{2.4}$$

Table 2.3 contains the values of the absolute viscosity for a range of temperatures. As shown in this table the absolute viscosity of water equals one centipoise at 20.2°C, which is about room temperature. This fact allows us to use the absolute viscosity of water as a relative standard for viscosities of other fluids. In comparison the absolute viscosity of air is about 0.17 cP and that of mercury about 1.7 cP.

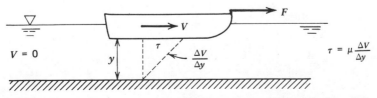

Figure 2.2 The interpretation of Newton's Law of Viscosity (Eq. 2.3)

Table 2.3
Absolute Viscosity of Water

Temperature (°C)	Absolute viscosity, μ (cP)
0	1.792
4	1.567
10	1.308
20	1.005
20.2	1.000
30	0.801
50	0.549
70	0.406
100	0.284
150	0.184

Example 2.4

A raft 3×6 m (9.84×19.68 ft) is dragged at a velocity of 1 m/s in a shallow channel 0.1 m deep measured between the raft and the channel bottom. Compute the necessary dragging force, assuming that Equation 2.3 is valid and that the temperature of water is 20°C.

Solution

From Equation 2.3

$$\tau = \mu \frac{\Delta v}{\Delta y} = \mu \frac{1 \text{ m/s}}{0.1 \text{ m}}$$

Assuming that the velocity changes linearly from zero at the channel to 1 m/s at the raft and from Table 2.3, μ at 20°C = 1.005 cP. By Equation 2.4 one obtains

$$\mu = 1.005 \times 10^{-3} \frac{N \cdot s}{m^2}$$

and by Equation 2.3

$$\tau = 1.005 \times 10^{-3} \frac{N \cdot s}{m^2} \frac{1.0 \text{ m/s}}{0.1 \text{ m}}$$

$$\tau = 1.005 \times 10^{-2} \text{ N/m}^2$$

The area of the bottom surface of the raft is

$$A = 3 \times 6 = 18 \text{ m}^2$$

The force required to pull the raft is the product of the viscous shear and the area, that is,

$$F = A \cdot \tau = 18(1.005)10^{-2} = 0.181 \text{ N} \ (0.04 \text{ lb})$$

Dividing the absolute viscosity by the density of the fluid at identical temperature results in the *kinematic viscosity, ν*. The concept of kinematic viscosity is often used by scientists and engineers. The unit of kinematic viscosity in the S.I. system is cm²/s, or the Stoke. For water, the value of the kinematic viscosity most commonly used in engineering computations is $\nu = 10^{-6}$ m²/s (1.55×10^{-3} in.²/sec).

Example 2.5
Determine the kinematic viscosity of water at 100°C.

Solution
From Table 2.2 the density of water at 100°C is

$$\rho = 958 \text{ kg/m}^3$$

From Table 2.3 the absolute viscosity of water at 100°C is

$$\mu = 0.284 \text{ cP}$$

By Equation 2.4 the latter term may be converted to

$$\mu = 2.84 \times 10^{-4} \text{ N} \cdot \text{s/m}^2$$

As the kinematic viscosity is defined as

$$\nu = \mu / \rho$$

one may write

$$\nu = \frac{2.84 \times 10^{-4}}{958} = 2.96 \times 10^{-7} \text{ m}^2/\text{s}$$

2.4 Surface Tension and Capillarity

Under normal conditions molecules of water bond in three dimensions. At the surface there is nothing to bond to in the upward direction and the surface molecules have some excess bonding energy they utilize in increasing their bonds along the surface, which results in *surface tension*. It is a layer of increased molecular attraction that, although only of the magnitude of one-millionth of a millimeter, has a significant influence on the physical behavior of water in a porous medium. Surface tension is therefore the added cohesion of water molecules on the surface. Its value depends on the temperature and electrolytic content of the water. For clear water at normal temperatures (about 20°C) the magnitude of the surface tension is 0.071 N/m (0.059 lb/in.).

Small amounts of electrolytes added to the water increase its surface tension. Salts dissolved from adjacent soil particles tend to increase the electrolytic content and, hence, the surface tension in groundwater. On the other hand, organic substances like soaps, alcohol, or acids decrease the surface tension. The reducing effect of soap on surface tension makes it possible to stretch water film while blowing soap bubbles.

Another important contributor to the physical effect of capillary rise is the *adhesion* of water to most solid materials. Solids that have positive adhesion to water are called *hydrophile* (water-liking), and those that repel water are *hydrophobe*. The latter have negative adhesion to water. Adhesion between fluids and solids is expressed by the contact angle at the edge of the contacting surfaces. Hydrophobe materials have a contact angle with water that is larger than 90 degrees. For example, the contact angle between water and paraffin is 107 degrees; hence, paraffin is a good waterproofing agent. Silver, on the other hand, is neutral to pure water; its contact angle is nearly 90 degrees. Quartz and other materials found in porous soils have a contact angle with water that is less than 90 degrees; this means that they wet well by water. The contact angle between ordinary glass and water containing impurities, for example, is about 25 degrees. In fact, the adhesive forces between water and soil particles are so large that they can be separated only by evaporating the water.

The capillary action—the rise of water in the small pores of soils and in thin glass tubes—is caused by the combined action of surface tension and adhesion. Figure 2.3 depicts the conditions present in a small diameter glass tube in which capillary rise of water takes place. By its adhesion to the solid wall water wants to cover as much solid surface as possible. However, by the effect of the surface tension the water molecules adhering to the solid surface are connected with a surface film in which the stresses cannot exceed the

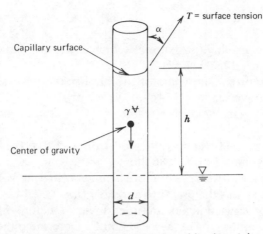

Figure 2.3 Capillary rise in a thin glass tube.

maximum possible surface tension of the water. The molecules in this surface film are joined to molecules below it by cohesive forces. As the adhesion drags the surface film upward, the film then raises a column of water filling the tube upward, against the force of gravity. The outcome of these factors is that the water in the small capillary tube, or in the small pores of soils, will rise upward against the force of gravity to a height at which the ultimate supporting capacity of the surface film is reached. Of course the column of water below the surface film is under tension, which means that the water pressure in a capillary tube is below the atmospheric pressure. It is as if the water molecules would be hanging off the surface film, held together by their molecular cohesion. The capillary rise is inversely proportional to the diameter of the tube or to the pore size in soils. Hence the finer the soil grains, the thicker the capillary layer in the soil mass. In the idealized case of a small diameter tube, the height of the capillary rise, h, is

$$h = \frac{4\sigma \cos \alpha}{d\gamma} \tag{2.5}$$

in which d represents the diameter of the tube, γ is the unit weight of the water, σ is its surface tension, and α is the contact angle representing the adhesion between the water and the tube. The angle α in this equation is usually assumed to be zero to 25 degrees for water in very small clean glass tubes; if the fluid is mercury, as in some manometers, the angle is 140 degrees.

Problems

2.1 Determine the absolute pressure at which water will vaporize at 60°C, at sea level. (*Ans.* 19,900 N/m²)

2.2 What is the approximate value of the atmospheric pressure at an elevation 1425 m above sea level? (*Ans.* 76 kN/m²)

2.3 Determine the density of sea water at 100°C temperature. (*Ans.* 996 kg/m³)

2.4 Two cubic meters of water are heated from 0 to 100°C. What will be the change in its volume? (*Ans.* 86 liter)

2.5 One cubic meter of water is compressed such that its volume is reduced by 0.01%. What was the pressure applied? (*Ans.* 2.14 × 10⁷ N/m²)

2.6 A 20 m² raft is dragged in a 0.2-m thick layer of water with a velocity of 2 m/s. The temperature of the water is 50°C. Determine the dragging force required to overcome the viscous shear. (*Ans.* 0.11 N)

2.7 Determine the kinematic viscosity of water at 0°C. (*Ans.* 1.79 × 10⁻³ m²/s)

2.8 The diameter of a clean glass pipe is 0.002 m. The pipe is filled with water at room temperature. Determine the capillary rise if the contact angle is 15 degrees. (*Ans.* 0.14 cm)

Chapter 3
Fluid
Statics

3.1 Hydrostatic Pressure

When a fluid is at rest, it is influenced by gravitational acceleration only. Regardless of their viscosity, all fluids behave the same way under static conditions. The reason for this is that viscous shear forces arise only when there is motion, according to Newton's law of viscosity, which was discussed in the previous chapter.

On the surface of a fluid in an open container, whether it is the surface of a glass of water or the surface of a lake, the only pressure that exists is the pressure of the air above it. Atmospheric pressure depends on the temperature and the elevation above sea level. Depending on meteorological conditions, the atmospheric pressure varies all the time. Weather reports include this information under the name of *barometric pressure*. It is reported in terms of inches of mercury at sea level. The barometer is a simple instrument consisting of a U-shaped tube with one end sealed and the other open to the outside air. The bottom of the tube is filled with mercury acting as the indicating fluid. With the sealed end of the tube protected from atmospheric pressure, the mercury behaves as a beam balance. The weight of the atmosphere at one end is indicated by the difference in height in the column of the mercury in the two connected tubes, as shown in Figure 3.1. The usual range of this barometric pressure is from 73 to 79 cm of mercury. Since the density of mercury is about fourteen times that of water, if one would use water in place of mercury in a barometer the water column would rise in its sealed part to a height of about ten meters. In hydraulic calculations the atmospheric pressure is standardized at sea level and normal temperature as 9.75 m (32 ft) of water. In everyday practice this is often simplified to ten meters. The pressure of the atmosphere is present everywhere on the earth's surface. In space, outside the atmospheric layer, a barometer would register zero pressure. When pressures are expressed in terms of barometric pressure, we speak of *absolute pressure*. In many hydraulic computations it is sufficient to

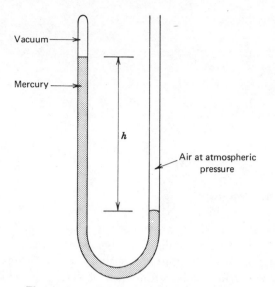

Figure 3.1 Simple mercury manometer.

refer to the surface of the water exposed to the atmosphere. Pressure related to open water surface is called *relative pressure* or gage pressure.

The magnitude of the hydrostatic pressure with respect to a water surface is expressed by the formula of

$$p = yg\rho = y\gamma \tag{3.1}$$

where the depth y is measured in the vertical downward direction from the surface corresponding to the gravitational acceleration g (Fig. 3.2). Using the notations introduced in Chapter 2, ρ is the density of the fluid and γ is the specific weight. As stated in the beginning of this section, pressures in fluids are scalar quantities; therefore at a point in the fluid they act in any and all directions equally. When the point considered is located on the surface of a solid immersed in or bounding the fluid, then the pressure is perpendicular to the solid surface. The reason for this is graphically explained in Figure 3.2; all nonperpendicular components of the now semispherical pressure distribution cancel out. Another way of explaining this concept is to consider the *hydrostatic force* acting on the surface. This force equals the hydrostatic pressure multiplied by the area on which it acts, or

$$dF = p \cdot dA \tag{3.2}$$

The elementary area dA is a vector quantity. To define an area we not only need to know its magnitude but also its orientation. Multiplying a scalar p and a vector dA, one obtains another vector dF. Also, since water cannot support shear forces without motion, the hydrostatic force must be perpendicular to the surface area on which it acts. Hence the orientation of the area dA in Equation 3.2 determines the direction of the elemental hydrostatic force dF that acts over it.

The determination of the total hydrostatic force F acting on a surface area A is done by summarizing all elementary force components acting on all small ΔA surface components. The resultant hydrostatic force could be considered as acting at a single point of the surface. This point on the surface is called the action point of the total hydrostatic force F.

Depending on the orientation of the surface area, computations of hydrostatic forces are more or less simple. In the case of a horizontal plane, the water pressure, which is constant throughout, is given by the product of the unit

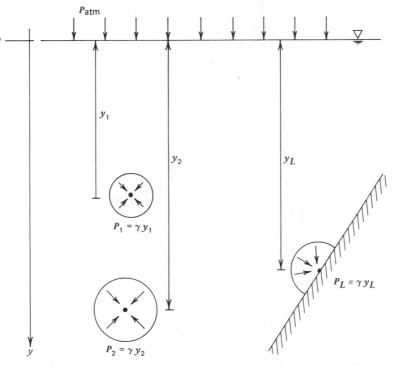

Figure 3.2 Hydrostatic pressure within the fluid and at the wall.

weight and its depth below the surface. The orientation of the force depends on which side of the surface the water is located. Figure 3.3 shows two such cases. In one case the water is above the surface; hence the force is acting downward. In the other case the water is below the surface; therefore the force acts upward. In both cases the point of application is at the centroid area. Another method giving the same result is to determine the pressure at the centroid of the area A and multiplying this pressure with A.

Example 3.1

The area of the bottom surface of a container shown in Figure 3.3a as A is 3.2 m² (34.4 ft²). The depth y is 4.1 m (13.45 ft). Determine the hydrostatic pressure acting at the surface and the total hydrostatic force.

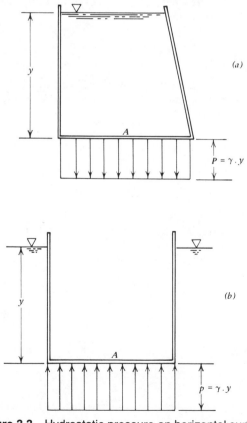

Figure 3.3 Hydrostatic pressure on horizontal surfaces.

Solution

Using Equation 3.1 to determine the hydrostatic pressure in water, we have

$$p = \gamma \cdot y = g \cdot \rho \cdot y$$

in which g, the gravitational acceleration, is 9.81 m/s^2, and ρ is the density of water, which equals 1000 kg/m^3. Accordingly, the specific weight γ is 9810 Newtons per cubic meter. Substituting one has

$$p = 9810(4.1) = 40{,}221 \text{ N/m}^2 = 40.221 \text{ kN/m}^2$$

This hydrostatic pressure is a scalar quantity, acting in all directions. In the case of Figure 3.3a where the water is inside the container, it is above the bottom and the pressure acts downward. The total hydrostatic force is computed by Equation 3.2 as

$$F = p \cdot A$$

in which A is 3.2 m^2. Accordingly,

$$F = 40.221(3.2) = 128.7 \text{ kN (29,000 lb)}$$

of force pointing downward. The location where this concentrated force acts is at the centroid of the area A.

3.2 Hydrostatic Force on Slanted Surfaces

On any surface subjected to a uniform hydrostatic pressure the resultant hydrostatic force is located at the centroid. On plane surfaces that are not parallel with the water surface the pressure varies according to the depth. The magnitude of the resultant hydrostatic force is still relatively easy to determine. It equals to the product of the surface area A and the hydrostatic pressure at the centroid of the area p_c. Hence the total force equals

$$F = A \cdot p_c = A \cdot \gamma \cdot y_c \tag{3.3}$$

where y_c is the depth of the water above the centroid of the area. The point of application of this total hydrostatic force is called the *center of pressure*. This point is located somewhat below the centroid at a distance e from it, measured along the steepest descent. The magnitude of e depends on the shape of the plane surface in question and on the depth of the centroid below the water surface. Before one can determine the location of the center of pressure, two important geometrical properties of areas must be known. These are the first and second moments of areas.

The *first moment of a plane area,* with respect to a coordinate line in the same plane, is computed by multiplying the magnitude of the area A with the perpendicular distance l between its centroid and the coordinate line, as shown in Figure 3.4; that is,

$$M = l \cdot A \tag{3.4}$$

The *second moment of an area I_0* is also called moment of inertia. If we are to determine I_0, the second moment of an area with respect to a coordinate line crossing its centroid, we have to sum the products of each small elemental surface component with their respective distances from the coordinate line. Generally this can be performed only by methods of integral calculus. However, for practical work it is sufficient to refer to tables showing values of I_0 for simple areas. Table 3.1 contains such information. For coordinate lines other than those crossing the centroid, second moments can be computed by

$$I_c = I_0 + Al^2 \tag{3.5}$$

where I_c is the moment of inertia with respect to the coordinate line, I_0 is the moment of inertia with respect to the parallel coordinate line crossing the centroid of the area, A is the area, and l is the perpendicular distance between the two coordinate lines. Figure 3.5 depicts these variables.

Once the concept of first and second moments of areas is known, we are ready to determine the location of the point where the resultant hydrostatic

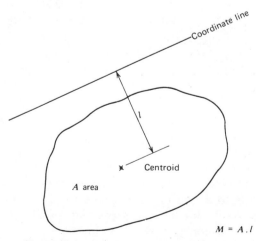

Figure 3.4 The first moment of an area.

Table 3.1
Location of Centroid, Area, and Moment of Inertia of Common Shapes

| Rectangle: | | $A = b \cdot h$ | $I_0 = b \cdot h^3/12$ |

Rectangle: $A = b \cdot h$ $I_0 = b \cdot h^3/12$

Triangle: $A = b \cdot h/2$ $I_0 = b \cdot h^3/36$

Circle: $A = \pi D^2/4$ $I_0 = \pi R^4/4$

Semicircle: $A = \pi R^2/2$ $I_0 = 0.110 R^4$

Ellipse: $A = \pi b \cdot h/4$ $I_0 = \pi b \cdot h^3/64$

Parabolic section: $y_0 = 3h/5$
$x_0 = 3b/8$ $A = 2b \cdot h/3$ $I_0 = \dfrac{8}{175} b \cdot h^3$

force acts on any known plane surface. If we refer to Figure 3.6 for our notations, the distance e between centroid and the hydrostatic force is given by

$$e = \frac{I_0}{lA} \tag{3.6}$$

where I_0 is the second moment of area A with respect to the centroid, l is the distance between the centroid and the line of intersection of the plane of A with the water level, and e is the distance between the centroid and the hydrostatic force, the center of pressure. For common geometric shapes found in hydraulic practice the center of pressure y_a is given in Table 3.2, where y_a is measured from the water surface.

Table 3.2

The Center of Pressure for Some Practical Hydraulic Problems. To Determine the Magnitude of the Force, Use Equation 3.3 with the Aid of Table. 3.1

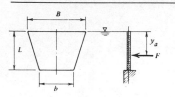

$$y_a = \frac{2B + 11b}{B + 2b} \cdot \frac{L}{6}$$

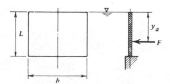

$$y_a = \tfrac{2}{3}L$$

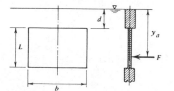

$$y_a = d + \frac{3d + 2L}{2d + L} \cdot \frac{L}{3}$$

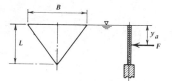

$$y_a = \frac{L}{2}$$

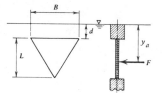

$$y_a = d + \frac{L(2d + L)}{2(3d + L)}$$

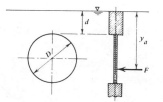

$$y_a = d + \frac{D}{8}\frac{8d + 5D}{2d + D}$$

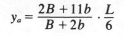

36

Table 3.2 *(Continued)*

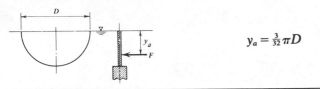

$$y_a = \tfrac{3}{32}\pi D$$

Example 3.2

A 1-m (3.28-ft) diameter flood gate placed vertically is 4 m (13.12 ft) below the water level at its highest point. See Figure E3.2. The gate is hinged at the top. Compute the required horizontal force to be applied at the bottom of the gate in order to open it. Assume that the pressure on the other side of the gate is atmospheric pressure and neglect the weight of the gate.

Solution

We know that the

$$\gamma \text{ of water} = 9.81 \text{ kN/m}^3$$

Figure 3.5 The interpretation of the amount of inertia with respect to the centroidal axis, I_o, and with respect to an arbitrary parallel axis at c distance from the centroid, I_c.

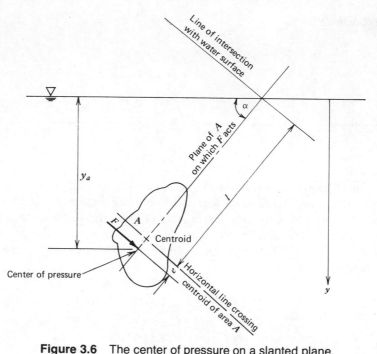

Figure 3.6 The center of pressure on a slanted plane.

$$\text{Area of the gate } = \frac{\pi}{4}(1)^2$$

$$A = 0.7855 \text{ m}^2$$

The depth of the centroid of the gate is $l = 4.5$ m. Next determine the location of F_1 acting on the gate.

From Equation 3.6,

$$e = \frac{I_0}{lA}$$

From Table 3.1,

$$I_0 = \pi R^4 / 4$$

$$= \pi(0.5)^4 / 4 = 0.0491 \text{ m}^4$$

Therefore,

$$e = \frac{0.0491}{4.5(0.7855)} = 0.0139 \text{ m}$$

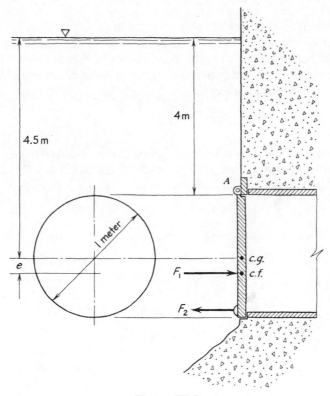

Figure E3.2

The force F_1 acting on the gate is located at a depth $y_a = (l+e)$. This equals $4.5 + 0.0139$, which equals

$$y_a = 4.5139 \text{ m}$$

The magnitude of this hydrostatic force is

$$F_1 = p_c A = 4.5 \times 9.81 \times 0.7855 = 34.67 \text{ kN}$$

To determine F_2 we take the moment of the two forces with respect to the hinge at A:

$$F_2 \times D = F_1 (R + e)$$

$$F_2 = \frac{34.67(0.5 + 0.0139)}{1}$$

$$= 17.82 \text{ kN (401 lb)}$$

Example 3.3

A rectangular gate is 4 m (13.1 ft) by 4 m set on a 45 degree plane with respect to the water surface. The centroid of the gate is located 10 m (32.8 ft) below the water surface, as shown in Figure E3.3. The gate is to be hinged in such a manner that there will be no moment rotating the gate under the given water depth. Locate the hinge shown in the figure.

Solution

The hinge should be a distance e below the centroid of the gate. First the terms included into the left-hand side of Equation 3.6 should be determined. Based on the notations shown in Figure E3.3, we have

$$l = 10/\sin 45° = 10/0.707 = 14.14 \text{ m}$$

The moment of inertia around the centroid, from Table 3.1, is

$$I_0 = b \cdot h^3/12 = 4.4^3/12 = 21.33 \text{ m}^4$$

The surface area of the gate is

$$A = b \cdot h = 4.4 = 16 \text{ m}^2$$

Substituting into Equation 3.6 we have

$$e = \frac{21.33}{14.14(16)} = 0.0943 \text{ m} = 9.43 \text{ cm}$$

Measuring from the edge of the water in the plane of the gate the distance of the hinge is $l + e = 14.14 + 0.0943 = 14.23$ m. From the top of the gate the hinge is at a distance of

$$l' = (h/2) + e = 2 + 0.09 = 2.09 \text{ m (6.86 ft)}$$

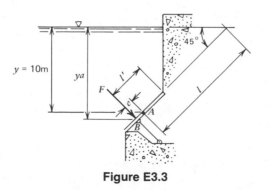

Figure E3.3

The total hydrostatic force there is, according to Equation 3.3,

$$F = 9.81(10)16 = 1569 \text{ kN } (390 \text{ kips})$$

Solving hydrostatic problems by a semigraphical approach is often advantageous. It involves the determination of the hydrostatic pressure at some critical points. Using these values pressure diagrams may be sketched by simple concepts of statics. In the case of slanted rectangles, for example, a pressure diagram may be constructed by computing pressures at the deepest and shallowest points of the area. As the pressure varies linearly with the depth, these two points may be connected by a straight line giving the pressure distribution (Fig. 3.7a). The same result is obtained if the pressure is

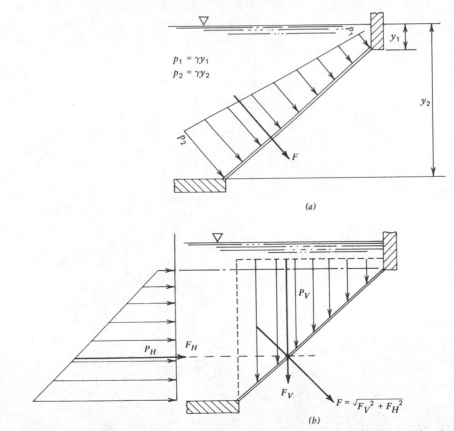

Figure 3.7 Semigraphical method for solving hydrostatic problems with slanted planes. (a) Direct solution. (b) Solving by horizontal and vertical pressure diagram components.

first separated into its horizontal and vertical components, as shown in Figure 3.7b. The vertical component of the hydrostatic force crosses the centroid of the pressure diagram, which, in turn, is composed of triangles and rectangles. In some cases it is convenient to break the pressure diagrams into such basic geometric components, resulting in several force components. The horizontal component of the pressure diagram may be handled in a similar manner. The vertical and horizontal force components are summed according to the fundamental concepts of statics. The intersection of the force components locates the position of the resultant hydrostatic force.

Examples 3.4 through 3.10

In the following examples, shown in Figures E3.4 through E3.10, a series of typical hydrostatic pressure diagrams are shown for plane surfaces. (Courtesy Ö. Starosolszky.) In each problem the shape of the vertical and horizontal pressure diagrams is shown along with the resulting force components.

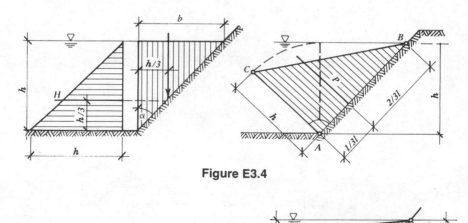

Figure E3.4

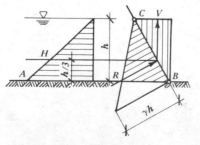

Figure E3.5

Figure E3.6

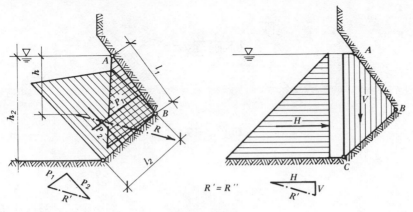

Figure E3.7

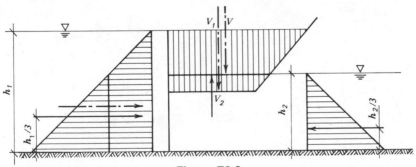

Figure E3.8

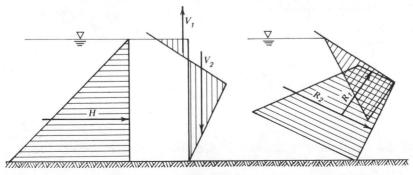

Figure E3.9

43

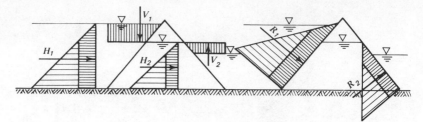

Figure E3.10

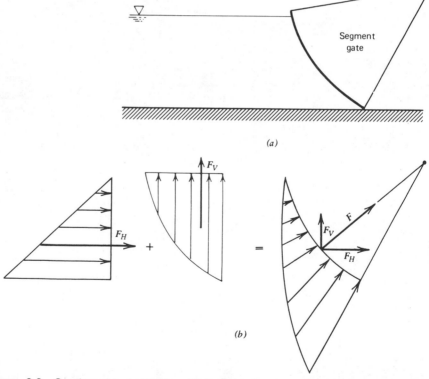

Figure 3.8 Semigraphical solution of hydrostatic pressure problem on a curved surface. (a) The geometry of the problem; (b) graphical addition of the pressure diagrams and the summation of the force vectors.

44

3.3 Hydrostatic Forces on Curved Surfaces

Hydrostatic forces acting on curved surfaces are somewhat more difficult to evaluate. In such cases a graphical approach to the computations is advantageous. Consider, for example, the segment gate shown in Figure 3.8a. As the normal lines of all surface elements cross at the pivot point of the gate—the center of the circle of which the gate is a segment—it is obvious that the resultant force will go through this pivot point also. Also note, however, that at each surface point the magnitude of the pressure as well as its direction varies. As such, it is convenient to separate the horizontal and vertical components of the pressure. This results, as shown in Figure 3.8b, in two rather simple pressure diagrams: a triangular one for the horizontal pressures and a segment-shaped one for vertical pressures. For the latter, depths below the water level at various points on the gate need to be multiplied by the unit weight of the water as a scaling factor in order to express vertical pressures. The vertical pressure component always equals the weight of water above the surface even in the case when the water is pushing the surface from below. To find the horizontal and vertical components of the resultant hydrostatic force, we only need to find the centroids of the two pressure diagrams through which the resultant must pass. By the laws of statics these component forces intersect at a point through which their resultant force passes. To find this point by graphical means is a simple matter, as is the graphical addition of these two components in order to obtain the resultant.

Examples 3.11 through 3.19

The following figures, E3.11 through E3.19, are graphical examples for the solution of hydrostatic force problems involving curved surfaces. (Courtesy of Ö. Starosolszky.) In each of the examples, the horizontal and vertical pressure diagrams are depicted.

Example 3.20

A 3 m (9.8 ft) diameter cylindrical (drum) gate is 8 m (26.2 ft) long. The water is on one side only, reaching the top of the gate. Determine the magnitude and direction of the resultant hydrostatic force acting on the gate.

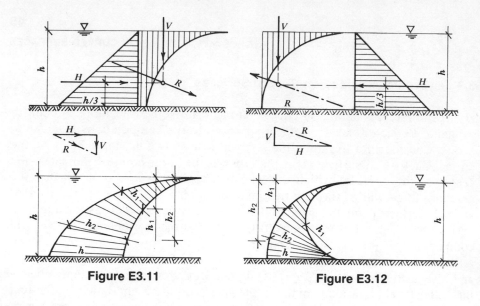

Figure E3.11

Figure E3.12

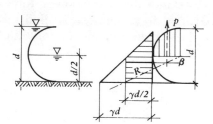

Figure E3.13

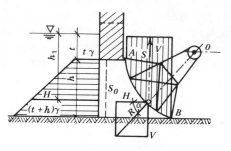

Figure E3.14

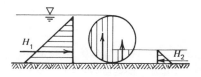

Figure E3.15

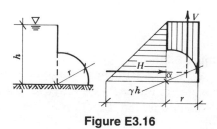

Figure E3.16

46

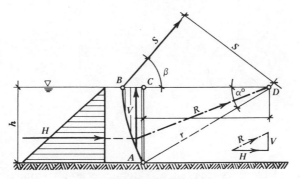

Figure E3.17

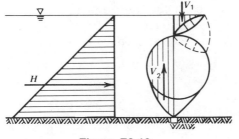

Figure E3.18

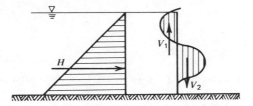

Figure E3.19

Solution

Separating vertical and horizontal pressure components the pressure diagrams are as shown in Figure E3.20. The horizontal force is acting at the centroid of the triangle, 2 m below the water surface. Its magnitude equals the area of the pressure triangle:

$$F_H = (\text{height})0.5p_1(\text{length}) = 3(0.5)9810(3)8 = 353 \text{ kN}$$

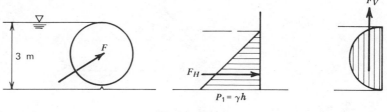

Figure E3.20

The vertical force is located a c distance away from the vertical centerline of the gate. The value of c may be determined from Table 3.1, in which

$$c = \frac{4R}{3\pi} = \frac{4(1.5)}{3\pi} = 0.63 \text{ m}$$

Its magnitude is the area of the semicircle with its diameter equal to p_1,

$$F_v = \gamma(0.5\pi R^2)(\text{length}) = 0.5\pi 1.5^2(8)9810 = 277 \text{ kN}$$

The resultant force is

$$F = \sqrt{F^2_H + F^2_V} = 448 \text{ kN (100 kips)}$$

The force acts through the intersection of its two components whose coordinates are 2 m vertical and 0.63 m horizontal with respect to the top of the water over the gate. The angle of the resultant force is

$$\tan \alpha = F_V/F_H = 0.78 \quad \text{and} \quad \alpha = 38.12 \text{ degrees}$$

3.4 Buoyancy and Stability

For floating objects the concepts developed in the previous paragraph apply also. As shown in Figure 3.9, horizontal force components on the two sides of the body cancel each other. The vertical components all but cancel each other except for the volume where the floating body displaces water, resulting in a force acting upward. Accordingly, this buoyant hydrostatic force on a floating body of any shape equals the volume of the body multiplied by the unit weight of water acting at the displaced mass center. This is the famous law of Archimedes. Of course, the floating body itself will also have its own weight, a force acting downward opposing the lifting force of the water. Whether a body will sink to the bottom, rise to the top, or experience an apparent weightlessness in water depends on whether the hydrostatic uplift is smaller, larger, or equal to the body's weight. This is the principle behind the operation of submarines.

Bodies floating on the water surface will submerge to a point at which the weight of the displaced water will equal their own weight. In order that such a floating object be in a stable condition, it is required that the mass center of the floating object be below the mass center of the displaced water. If the two mass centers coincide, the object will be unstable. Overloaded boats capsize because their mass center, including that of the load, is higher than that of the displaced water.

Example 3.21

What would the draft of a 4 m by 8 m rectangular barge be if its weight is 156 kN?

Solution

The volume of the displaced water is

$$\text{width} \times \text{length} \times \text{draft} = 4 \times 8 \times d = 32 \times d$$

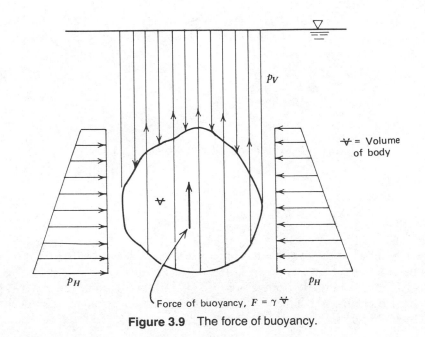

Figure 3.9 The force of buoyancy.

The weight of the displaced water must equal the weight of the barge; therefore,

$$32 \times d \times \gamma = 156 \times 10^3$$

and, from here

$$d = \frac{156 \times 10^3}{32 \times 9810} = 0.5 \text{ m}$$

Example 3.22

One-seventh of an iceberg's mass is above the surface of the ocean. Determine the specific weight of the seawater.

Solution

If the weight of the iceberg is W, then, by the law of Archimedes,

$$W = V\gamma_{\text{seawater}}$$

in which V is the volume of seawater displaced by the iceberg. If the volume of the iceberg is V', then its weight is

$$W = V'\gamma_{\text{ice}}$$

since ice, from Table 2.2, has a specific weight of 8996 N/m³. Since one-seventh of the iceberg is above the ocean surface, we can write

$$V(1 + \frac{1}{7}) = 1.143V = V'$$

and by substituting we have

$$W = 1.143V(8996) = V\gamma_{\text{seawater}}$$

By dividing through with V we obtain

$$\gamma_{\text{seawater}} = 1.143 (8996) = 10.0 \text{ kN/m}^3 \ (63.7 \text{ lb/ft}^3)$$

Problems

3.1 How much is the hydrostatic force in Example 3.1 if it is to be expressed in terms of absolute pressure? (*Ans.* 136 kN/m²)

3.2 In Figure 3.3*b* the bottom area is 5.2 m² and the depth of the water is 4.5 m. Determine the magnitude, direction, and location of the resultant hydrostatic force. (*Ans.* 230 kN)

3.3 On a closed vertical sluice gate that is 2 m wide, the water level is 3 m from the bottom at one side and 1.5 m deep on the other side. What is the resultant hydrostatic force on the gate and what is its location? (*Ans.* 66 kN, 1.17 m)

3.4 A 45 degree slope contains a closed manhole cover of 1.2 m diameter. The center of the manhole cover is under 2.4 m of water. Determine the magnitude and location of the hydrostatic force acting due to the covering water. (*Ans.* 26.6 kN)

3.5 Determine the second moments of a rectangular area that is 4 m wide on each side with respect to the centroidal axis and a parallel axis 3.4 m to the side. (*Ans.* 487.8 m^4)

3.6 A vertically placed circular flood gate is 1 m in diameter. What should be the moment at its upper hinge to open the gate against the pressure of 3 m of water measured from the top of the gate? (*Ans.* 14 kN · m)

3.7 In Example 3.14 the diameter of the drux gate is 5 m and its length is 9 m. Determine the magnitude and location of the resultant hydrostatic force if there is no downstream water and the upstream water level reaches the top of the gate. (*Ans.* 2056 kN)

3.8 Repeat Problem 3.7 with an additional load resulting from a downstream water level of 1.3 m.

3.9 A 2.5 m by 6 m pontoon is 1 m deep. Its dry weight is 12 kN. How much dry sand (sp. wt. 25.5 kN/m^3) can be put into the hold before the pontoon sinks? (*Ans.* 5.3 m^3)

Chapter 4
Fluid
Dynamics

4.1. Description of A Moving Fluid

In attempting to describe motion in fluids one may take either of two approaches: Follow the path of a fluid particle through space defining the location and the time of arrival at each point. Or, select points in space and note the magnitude and direction of the velocity of fluid particles as they pass through the point selected.

By *velocity* we mean the change of position of a water particle in the moving fluid during a certain time interval. In Figure 4.1, for example, a particle may be found at a position x_1 at time t_1. If the same particle at a later time t_2 is found at position x_2, then the velocity of the particle is $(x_2-x_1)/(t_2-t_1)$. As we take shorter and shorter time intervals (t_2-t_1), the path of the particle between the end points x_1 and x_2 will be less and less distinguishable from the straight line connecting the end points. For an extremely small interval (Δt), the line becomes tangent to the path at point x_1. In this case we speak of a velocity vector at point x_1. The direction of a velocity vector shows which way the particle is moving at this point. The magnitude of the velocity vector signifies how fast the particle is moving. Now, if we are looking at the point x_1 as a location fixed in space through which a stream of fluid particles passes, we can define another important principle: If the succession of water particles passing through x_1 has identical velocity vectors, we speak of *steady flow*. Conversely, if the velocity vector changes for successive fluid particles, we have *unsteady flow*.

Acceleration of a fluid means a change in velocity. There are two kinds of accelerations since the velocity can change in place as well as in time. In a pipe of enlarging diameter, the velocity of the flow decreases as it passes from a section of small diameter to a section of large diameter. This form of deceleration, or negative acceleration, is called *spatial acceleration* since the change of velocity occurs in space. If the discharge flowing through a certain cross-sectional area varies with time, we speak of *temporal acceleration*. For

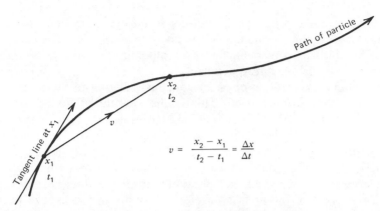

Figure 4.1 Definition of velocity.

almost all topics considered in this book we will assume that no temporal accelerations exist. This means that the flow is steady. Internal accelerations within the fluid result from the relative movement of the water because of external forces. By relative movement we mean the motion of water with respect to its surroundings. The external forces may be created by the gravitational acceleration g, which acts everywhere on earth. The driving force, in this case, comes about by virtue of Newton's law of motion. If the only acceleration acting on the fluid particles is the one due to gravity and the water is contained, we have no cause for relative motion between the fluid particles. As long as the water is contained, it will be at rest. Forces caused by water at rest are in the domain of *hydrostatics,* discussed in Chapter 3. Forces caused within the moving mass of fluid, on the other hand, are within the domain of *hydrodynamics,* which is the study of the behavior of water in motion. Its basic physical laws are Newton's laws, the laws of conservation and the laws of energy transfer.

In the science of fluid mechanics we distinguish between compressible and incompressible fluids. Laws relating to the behavior of compressible fluids like air, steam, or gases fall within the domain of aerodynamics, or gas dynamics. Because for all pressures encountered in hydraulic applications, water in its liquid form retains a constant volume for a given amount of mass, it is considered to be incompressible. The assumption of incompressibility simplifies the fundamental fluid mechanics principles. It allows us to measure amounts of water in "volumetric" terms instead of in mass terms.

Discharge means the volume of water flowing through a certain cross-sectional area during a specified time period. It is usually measured in cubic meters per second. In some specific applications other discharge units may be found. For example, discharge of pumps is quoted in gallons per minute, output of water treatment plants may be given in million gallons per day, and so forth. A table of conversion factors including the most frequently used discharge units is included in Appendix 1.

In the explanation of some physical concepts relating to the flow of fluids, the term *mass flow rate* is used. This is the amount of fluid mass passing through a given area during a specified length of time. To obtain the mass flow rate, discharge is multipled with the density of the fluid. The *dimension of mass flow rate in the S.I. system is kilograms per second (kg/s).*

In practice one often finds that as a discharge Q flows through a cross-sectional area A of a flow channel, the velocity of the fluid particles flowing through each point of the area is different. Usually in the central portion of the cross section we find the highest velocities, while along the edges the velocity may be almost zero. It is convenient in such cases to define the *average velocity* as

$$v_{\text{average}} = Q / A \tag{4.1}$$

where A is considered perpendicular to the direction of the flow, as shown in Figure 4.2. In some topics, such as the measurement of flow by velocity meters, when we speak of velocity we refer to a particular velocity vector in a given point of the flow field.

Example 4.1

In a 50 cm (19.7 in) diameter pipe water flows with a conical velocity distribution (see Fig. E4.1). At the pipe walls the velocity is zero, and at

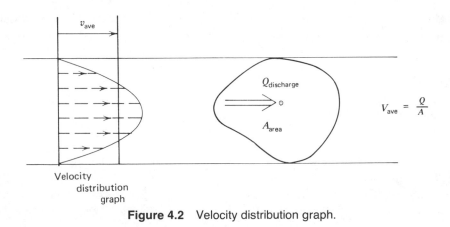

Figure 4.2   Velocity distribution graph.

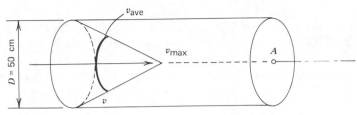

Figure E4.1 Conical velocity distribution in a pipe.

the center its maximum value is 2.5 m/s. Determine the average velocity, the discharge, and the mass flow rate.

Solution

The cross-sectional area of the pipe is

$$A = \frac{\pi D^2}{4} = 0.2 \text{ m}^2$$

To obtain the average velocity first the volume of a conical velocity distribution function is computed. The volume of the 2.5 m/s high cone is

$$\text{Vol}_{\text{cone}} = \tfrac{1}{3}(\text{base area}) \cdot (\text{height}) = \tfrac{1}{3}(0.2)2.5 = 0.17$$

The height of a cylinder of equal volume is the measure of the average velocity:

$$\text{Vol}_{\text{cylinder}} = (\text{base area}) \cdot (\text{height}) = 0.2(v_{\text{ave}}) = \text{Vol}_{\text{cone}}$$

and, from here,

$$v_{\text{ave}} = \frac{0.17}{0.2} = 0.85 \text{ m/s}$$

The discharge is given by Equation 4.1 as

$$Q = A \cdot v_{\text{ave}} = 0.2(0.85) = 0.17 \text{ m}^3/\text{s}$$

The mass flow rate is the product of the discharge and the density. The density may be taken as 1000 kg/m³, hence

$$\text{mass flow rate} = 1000(0.17) = 170 \text{ kg/s}$$

Example 4.2

The discharge of a creek was measured to be 0.5 m³/s (17.7 ft³/sec) at 9 a.m. and 0.6 m³/s (21.2 ft³/sec) at 4 p.m. on the same day. The area of the flow was 1.445 m² (15.55 ft²) during the first measurement and 1.57 during the second measurement. Determine the average velocities of the flow and the average temporal acceleration during this time period.

Solution

At 9 a.m. the velocity was

$$v_1 = \frac{Q_1}{A_1} = \frac{0.5}{1.445} = 0.346 \text{ m/s}$$

At 4 p.m.

$$v_2 = \frac{Q_2}{A_2} = \frac{0.6}{1.57} = 0.382 \text{ m/s}$$

The time interval between 9 a.m. and 4 p.m. is 7 hours, or

$$T = 7 \times 3600 = 25,200 \text{ seconds}$$

The average temporal acceleration

$$a = \frac{v_2 - v_1}{T} = \frac{0.382 - 0.346}{25,200} = \frac{0.036}{25,200}$$

$$= 1.43 \times 10^{-6} \text{ m/s}^2 \ (4.7 \times 10^{-6} \text{ ft/sec}^2)$$

4.2 The Conservation of Mass

The conservation of mass is one of the three conservation laws of physics. It states that mass cannot be created nor destroyed. This concept gives rise to the *equation of continuity,* which states that within any hydraulic system the discharge flowing in, the stored volume within, and the discharge flowing out, must be balanced; in other words, all volumetric quantities must be accounted for. Since we consider water to be incompressible, mass or volume may be used interchangeably in this respect. To put the concept into mathematical form we may write our continuity equation as

$$Q_{in} - Q_{out} = \text{change in storage} \qquad (4.2)$$

The equation above is often used in the analysis of reservoirs and in routing of floods in rivers. It is important that the time scale be the same on both sides of the equation. For example, if the discharges are available in the units of cubic meters per second and the change in storage is desired for periods of, say, six hours, an appropriate conversion factor must be introduced.

In the case when no change in storage is possible, as in a pipe flowing full, the right-hand side of Equation 4.2 will reduce to zero, which means that what goes in, must come out. Figure 4.3 illustrates these concepts. In certain applications, as in rivers or complex pipelines, the problem is broken into smaller components joined together at certain points. In this case, care should be exercised in selecting the proper sign for each discharge component. A common sign convention is to consider discharges flowing into the hydraulic component as positive and outflowing discharges as negative. At points where two hydraulic components are joined, the sign of the discharge changes.

Example 4.3

A 30-m (98.4-ft) wide river section is 2000 m (6560 ft) long. The inflow discharge is 45 m³/s (1588 ft³/sec) and the outflow discharge is 30 m³/s (1059 ft³/sec). How much will be the change of the average water level in the stream if the given flow conditions are maintained for a period of three hours?

Solution

By Equation 4.2 the change of storage is

$$45 - 30 = 15 \text{ m}^3/\text{s}$$

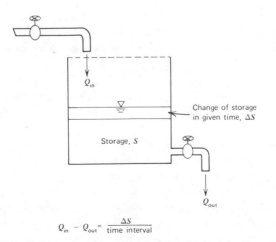

$$Q_{in} - Q_{out} = \frac{\Delta S}{\text{time interval}}$$

Figure 4.3 Explanation for the principle of conservation of mass.

This will be the amount that goes into storage on the top of the amount of water already in the channel section. The surface area of the river in the channel section studied is

$$A = 30 \text{ m} \times 2000 \text{ m} = 60,000 \text{ m}^2$$

For the three-hour period the total volume that is to be stored is

$$3(3600)15 = 162,000 \text{ m}^3$$

The rise in the water level will be

$$\frac{162,000}{60,000} = 2.7 \text{ m (8.9 ft) or 0.9 m/hr}$$

4.3   Work, Energy, and Power

To create and maintain motion in a fluid *work* must be performed. There are many ways to describe work. For instance, if a force F is required to push water through a pipe of length L, then the work done is

$$\text{work} = W = F \cdot L = \text{Force} \times \text{Distance} \qquad (4.3)$$

Another way of expressing work is by the *power* required to do it. Power is the rate of doing work. Accordingly, if a pump requires P power over a T time period to do the work required, then

$$\text{work} = W = P \cdot T = \text{Power} \times \text{Time} \qquad (4.4)$$

Yet another way of expressing work done is by referring to the *energy* used in performing it. The difference between the energy available initially and the energy remaining after the work is done equals the work done, that is,

$$\text{work} = W = \text{Energy}_{\text{in}} - \text{Energy}_{\text{out}} \qquad (4.5)$$

Based on the various ways work may be described, its dimensions may be written either in Newton-meters, kilowatt-hours, or in the many other ways energy may be defined.

From Equation 4.4 it is clear how closely the concept of power is related to that of work. To perform a certain amount of work in a shorter period requires more power than to perform the same work over a longer period of time. Chapter 7 of this book discusses this concept in greater detail in connection with power required to drive pumps. Hydraulic power generated by turbines is another matter dealing with this subject. When considering electric power needs to drive pumps, or the electric power generating capacity of a turbine in a dam, the power is usually expressed in terms of watts (W), or, more practically, kilowatts (kW). Watt is defined as one Newton · meter/ second. To move a discharge Q against a pressure of γH the required power is

$$P \text{ (kW)} = \frac{\gamma \cdot H \cdot Q}{1000} \tag{4.6}$$

where γ is in N/m³, Q is in m³/s, and H is in meters. Mechanical energy is often expressed in horsepower, hp. One horsepower is 75 kilowatts.

The ability to do work is called *energy*. Energy therefore is simply stored work. In fluid mechanics, energy is most often expressed as energy per unit mass of fluid. There are various forms in which energy appears in nature; put another way, one can store work in several ways. Temperature indicates the amount of *heat energy* a substance contains. If we know the specific heat of a substance, we may compute the amount of heat energy that could be removed from it by reducing its temperature. We learned in Chapter 2 that *temperature* and *pressure* represent internal energy in matter. A lump of coal has *chemical energy* that may be transformed into heat energy by the chemical reaction called burning. An object placed higher in a gravitational field has more energy by virtue of its relative elevation than the same object at a lower elevation. This is called *potential energy*. A compressed spring or a stretched strip of rubber has *elastic energy*. The force applied to them gives rise to *pressure energy*. Pressure energy and elastic energy may be illustrated by considering an inflated bicycle tube: The elastic rubber tube is stretched, gaining elastic energy; since the air in the tube is under pressure, it is compressed and has pressure energy with respect to the surrounding air, which itself is under pressure by the atmosphere. Moving objects contain *kinetic energy*. Their velocity may propel them to higher elevation, like water from a fire hose, in which case the kinetic energy is converted into potential energy.

As we have seen above, some energies can be freely converted among each other; others are not so reversible. Elevation, or potential energy, pressure energy, and kinetic energy are reversible energies. These are called *hydraulic energies*. On the other hand, heat energy cannot be harnessed by hydraulic means. Once a portion of our hydraulic energy converts into heat through viscous shear action, it is lost to us; therefore, we refer to it as energy loss. The heat energy converted from sunshine causes the water from the oceans to evaporate, keeping the water in constant circulation through the hydrologic cycle. Therefore the ultimate source of hydraulic energies in nature is the sun. The mathematical formulas for the three hydraulic energies expressed for one kilogram of fluid flowing are as follows:

$$\text{Kinetic energy} = v^2 / 2g$$

$$\text{Pressure energy} = p / \gamma \qquad (4.7)$$

$$\text{Gravitational (potential) energy} = z$$

where z is an elevation over a conveniently selected datum plane.

4.4 The Conservation of Energy

An important concept of physics is the law of energy conservation. It states that energy cannot be lost, though it may be converted into other forms. In other words, the theorem states that in a hydraulic system the sum of all energies is a constant. Writing this total energy in mathematical form,

$$E = \frac{v^2}{2g} + \frac{p}{\gamma} + y = \text{const} \qquad (4.8)$$

is called *Bernoulli's equation* after one of the famous hydraulicians. Equation 4.8 includes only hydraulic energies. It does not include the heat energy term. Heat energy is always present although the science of hydraulics is not concerned directly with it. Only the change in the heat energy appears in the equations. Heat energy changes appear as water flows from one point to another in the form of viscous shear losses, taking away some portion of the total hydraulic energy E. For this reason, E is called the *total available hydraulic energy* at any point considered. As this total available energy decreases along the direction of the flowing water between two points, as shown in Figure 4.4, the law of energy conservation has this form:

$$E_{\text{in}} = E_{\text{out}} + \Delta E$$

$$\frac{v1^2}{2g} + \frac{p_1}{\gamma} + y_1 = \frac{v2^2}{2g} + \frac{p_2}{\gamma} + y_2 + h_L \qquad (4.9)$$

where subscripts 1 and 2 refer to points shown in the flow field and the term h_L refers to the energy lost in the form of heat. This lost heat energy is actually still present in the environment in the form of an imperceptible temperature rise that is generally dissipated in the surrounding atmosphere. The h_L term is commonly called head loss because it is the apparent reduction of the height of the total available hydraulic energy E as the water moved from one point to the other.

The terms in Equation 4.8 all have the dimension of elevation, measured above a horizontal reference level, called *datum plane,* which is perpendicular to the direction of the gravitational acceleration. The sum of the last two

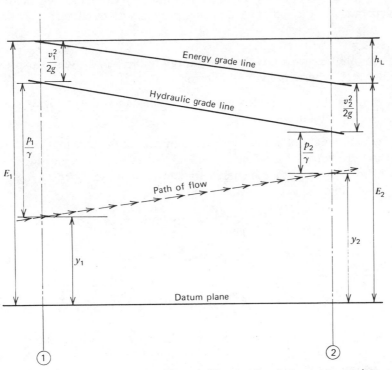

Figure 4.4   Interpretation of Bernoulli's equation between two points.

terms, $p/\gamma + y$, is sometimes called the piezometric height. This is the height to which water would rise in a pipe with one of its ends inserted into an arbitrary point of the flow field. The line, shown in Figure 4.4, connecting several points of such piezometric measurements along the path of flow is called *hydraulic grade line*. It is always below the total *energy grade line* by an amount equal to the kinetic energy head, $v^2/2g$, at the point of the piezometric measurement. In open channels, when the water surface is exposed to the atmosphere and hence is not under pressure, the hydraulic grade line coincides with the water surface.

Example 4.4

A mountain creek discharges 3 m³/s (105.9 ft³/sec) along a rocky channel without apparent freezing in subzero weather. The channel elevation is reduced by a height of 80 m (262.4 ft) over a channel length of 1 km (0.6.

mile). The flow in the channel is uniform. Compute the heat energy generated by the moving water that prevents its freezing.

Solution

From physics we know that the energy in the S.I. system is expressed in kilowatts, where, in hydraulic terms,

$$1 \text{ kW} = \frac{\gamma QH}{1000} = \frac{(\text{unit weight, N/m}^3)(\text{discharge, m}^3/\text{s})(\text{head loss, m})}{1000}$$

The unit weight of water, expressed in S.I., is 9810 N/m³. The discharge in the problem is 3 m³/s, and the head loss is 80 m. The head loss may be looked upon as potential energy transformed in the process. Hence, after substituting, we get

$$\frac{9810(3)80}{1000} = 2354 \text{ kW (3159 hp)}$$

Distributing this total energy over the channel length in a uniform manner, we have

$$\frac{2354 \text{ kW}}{1000 \text{ m}} = 2.35 \text{ kW/m}$$

which is equivalent to the power used by over twenty-three 100-Watt light bulbs for each meter of channel length. This is the heat energy that keeps the water from freezing.

Example 4.5

A 1000-m (3280-ft) long pipe is 30 m (98.4 ft) higher at the entrance point and 10 m higher at the exit point than the reference level (Fig. E4.5). The pipe diameter is constant. The velocity in the pipe is 8 m/s. The water elevation at the entrance is 12 m above the pipe. The pressure at the exit point is the atmospheric pressure. Calculate the energy loss due to the flow in the pipe.

Solution

The pipe has a constant diameter. The kinetic energy is

$$\frac{v^2}{2g} = \frac{v^2{}_1}{2g} = \frac{v^2{}_2}{2g}$$

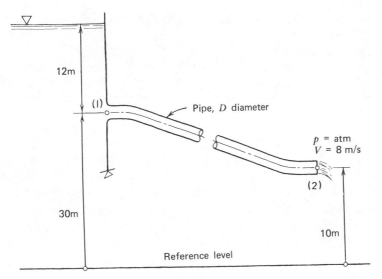

Figure E4.5

$$= \frac{(8)^2}{2 \times 9.81} = 3.26 \text{ m}$$

From Equation 4.9, if we take the points at the entrance and the exit point as (1) and (2) we have, therefore,

$$\frac{v^2_1}{2g} + \frac{p_1}{\gamma} + y_1 = \frac{v^2_2}{2g} + \frac{p_2}{\gamma} + y_2 + h_L$$

$$3.26 + 12 + 30 = 3.26 + 0 + 10 + h_L$$

Hence, $h_L = 32$ m (105 ft) = hydraulic energy loss in the pipe due to friction and other factors.

4.5 Force and Momentum

The fundamental equation of mechanics, Newton's second law of motion, can be arranged in terms of the common variables of fluid mechanics by writing

$$F = m \cdot a = (\rho Q \, \Delta t)(\Delta v / \Delta t) \qquad (4.10)$$

where m is the mass that can be written in terms of mass flow rate, and a is the acceleration, which may be written in terms of the time rate of change of velocity. By rearranging one gets

$$F = \Delta(\rho Q v) = \Delta(\rho A v^2) \tag{4.11}$$

where the term in parenthesis is called *momentum*. The momentum can be considered a property of the flowing fluid just like the kinetic energy. In words, Equation 4.11 states that the time rate of change of momentum results in a force. Conversely, one may state that the resultant of forces acting on a certain body of fluid causes a change in its momentum. One may be reminded here that the "time rate of change" is a concept that was also used in connection with the definition of acceleration. There we noted two things that may change in the condition of flow with time: it could change spatially, or it could change with time, or both. Assuming steady flow the velocity and the discharge will not change in time, but the direction of the velocity may still change in space. Therefore there is a change of momentum resulting in a force.

For practical applications the first step of the analysis is to define the body of the fluid that is being considered. This portion of the space occupied by the fluid is called a *control volume*. Surfaces of a control volume are either perpendicular or tangential to the velocities of the flow. Tangential surfaces are either solid boundaries within which the flow takes place, such as the internal surfaces of a pipe section, or the free surface of the water flowing in a channel. The other surfaces of the control volume are perpendicular, or "normal," to the velocities of the flow as it enters or leaves the control volume. The control volume is permanently fixed in space and time. As an example Figure 4.5 shows the control volume associated with a constricting pipe elbow. The concept is identical to the free body diagram known in statics, where components removed are replaced by equivalent forces. Once the control volume is well defined, the next step is to enumerate and show all forces, pressures, and velocities that are present and acting within or on the surfaces of the control volume. These are the velocities entering and leaving the control volume, the pressures, or forces acting on the surfaces normal to the flow, the shear forces and normal forces (or stresses) along the solid boundaries, and the gravitational and other body forces acting on the fluid enclosed. We must keep in mind that the force F in Equation 4.11 is the resultant of all forces present. The resultant of the active forces is countered by an equal but opposite reaction force. The force of resistance is, for example, the force exerted by the walls of the pipe elbow on the fluid. In many applications the problem is simplified by the fact that several of the forces present may be of negligible influence. For example, in the case of our

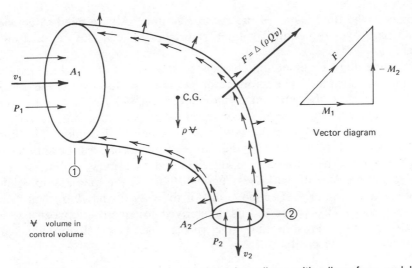

Figure 4.5 Control volume of a constricting pipe elbow with all surface and body forces shown.

constricting pipe elbow shown, the friction forces acting between the fluid and the pipe wall are negligible. Also negligible is the weight of the fluid, the portion of the body force due to gravitational acceleration, and the resulting small difference of the normal surface pressures exerted by the fluid on the pipe walls, and, of course, the resultant of these normal pressures itself, when contrasted by the force generated by the change of momentum due to the flow.

In order to obtain the force F shown in Figure 4.5 the momentum of the flow at both inlet and outlet must be evaluated. Equation 4.11 may be applied, giving

$$M_1 = \rho\, Q \cdot v_1 \qquad (4.12)$$

and

$$M_2 = \rho\, Q \cdot v_2$$

$$F = M_1 - M_2$$

Plotting these values as a vector diagram (shown in Figure 4.5) will result in the force F sought.

The advantage of interpreting Newton's law of motion in the form of momentum changes is that we do not have to concern ourselves with what happens to the fluid inside of the control volume. All we need to consider is the changes over the control volume's surfaces in terms of momentums and forces.

The force due to momentum changes on the blades of a turbine may cause the turbine to rotate. The product of an impulse force and the radius of rotation results in the moment of momentum or, as it is called, the torque. The concept of torque, explained graphically in Figure 4.6, is applied in Equation 7.3 to determine the power requirements of pumps. The impulse force transferred to the water by the ship's propeller causes the ship to move. This force must be sufficient to overcome the various losses due to this relative motion. One of the losses, among others, is the momentum needed to be given to the water first to move out of the way of the ship, then to move back behind it. This process causes relative movements between the water particles around the ship that give rise to additional losses due to the resistance by the viscosity of the water.

4.6   The Conservation of Momentum

Another conservation law of physics is the *conservation of momentum*. Based on Equation 4.11 the change of momentum of a given body of fluid may be written as

$$M_{out} = F + M_{in}$$

$$\rho Q v_2 = F + \rho Q v_1 \tag{4.13}$$

where subscripts 1 and 2 refer to the inlet and exit conditions of the velocity vector on the control volume and F is the force acting on the control volume. Since both velocities and forces are vector quantities, it is advantageous to solve such problems in a scalar form after separating the forces and momentums into their x, y, and z components of a rectangular coordinate system. Equation 4.13 embodies the *law of momentum conservation*, which states that momentum may not be lost in a hydraulic system, although some of it may be converted into impulse forces. Hence, if the mass flow rate and the physical configuration of the flow channel causing a change in flow direction are known, the resulting impulse forces acting on the hydraulic component or structure may be computed.

Another example of the use of conservation of momentum theorem may be the behavior of a jet of water in air. Let us consider a portion of a jet as a control volume with a constant Q discharge flowing with a velocity v. A distance away from the nozzle the jet tends to break up because of the difference of velocity across it. The difference in velocity is generally caused

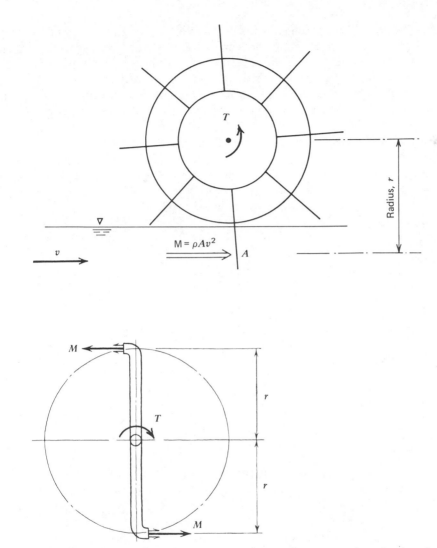

Figure 4.6 The moment of momentum, or torque, acting on a water wheel and on a rotating garden sprinkler.

by the retarding friction in the nozzle and the air resistance along the boundary of the jet. Once the jet breaks up, it will have to carry the air within its separated portions. To do this, the air mass carried will take some of the momentum away from the water. In other words, some air will tend to share the original momentum content of the water. But since the total momentum must remain the same, the velocity of the water will have to be reduced, as the discharge will not change. (Another way of looking at this problem is to consider the density decreasing because of the mixing of air and water and the discharge increasing by the same effect.) A considerably more influential force on the jet is the resistance of the air against it. The air resistance is proportional to the square of the velocity and the cross-sectional area of the jet, or $v^2 \cdot a$. Another force acting is a body force, the gravitational acceleration multiplied with the mass of fluid flowing. Applying these forces upon the control volume of the jet will result in a change in its momentum. Since the mass flow rate is considered the same (with the mixing of the air neglected), the velocity will have to undergo a change. This change is of course a gradual process, resulting in the well-known parabolic path of a jet stream.

Example 4.6

An 8 cm (3.15 in.) diameter 90 degree pipe bend carries 0.05 m³/s (1.77 ft³/sec) of water. Determine the magnitude and direction of the force resulting from the change of momentum of the flow.

Solution

The density of the water is 1000 kg/m³. The cross-sectional area is

$$A = \frac{\pi D^2}{4} = 0.005 \text{ m}^2$$

The velocity is

$$v = Q/A = 0.05/0.005 = 10 \text{ m/s}$$

The only change in the momentum as the water flows through the pipe bend will be a 90 degree change in direction. The magnitude, if the momentum will be the same throughout, is $M = \rho \cdot Q \cdot v = 1000$ (0.05)10 = 500kg $\cdot$ m/s² = 500 N. This force vector will act in the direction perpendicular to the inlet, as well as the outlet, of the pipe. The angle of the pipe bend being a right angle, the resultant force vector of the vector diagram may be computed by either the theorem of Pythagoras, or by the cosine of the angle between the momentum and resultant force vector. By the latter,

$$\cos \delta = (M_{in}/F) = \cos 45° = 0.707$$

and from here, since M is known,

$$\frac{500}{0.707} = F = 707 \text{ N (159 lb)}$$

Example 4.7

Assume that in the previous example the inlet diameter of the pipe remains the same but the outlet diameter will be 12 cm (4.7 in.). Recompute the problem.

Solution

The area of the exit will be

$$A_2 = \frac{\pi 12^2}{4} = 0.01 \text{ m}^2$$

Hence the velocity at the exit will be

$$v_2 = Q/A_2 = 0.05/0.01 = 5 \text{ m/s}$$

The momentum at the exit will be

$$M_2 = 1000(0.05)5 = 250 \text{ N}$$

To determine the magnitude of the resultant force F one may use the Pythagoras formula,

$$F = \sqrt{500^2 + 250^2} = 559 \text{ N (125.8 lb)}$$

The angle δ between the momentum at the inlet and F is

$$\cos \delta = 500/559 = 0.89$$

Therefore the angle δ sought is 26.5 degrees.

4.7 The Transformation and Loss of Energy

The science of hydraulics is primarily concerned with the determination of the magnitude of the energy loss, h_L, under various circumstances. In most practical applications, as in the flow of water in pipes and open channels, the energy loss is assumed to be proportional to the square of the average velocity of the flow, or, in other words, to the kinetic energy. Seepage is one exception, where the energy loss is considered to be proportional to the velocity. This concept of proportionality gives rise to coefficients or *factors of proportionality*. This is why hydraulics is a science of experimentally determined coefficients. Formulas for pipe flow include a coefficient f, which is called the friction factor; open channel flow formulas include a coefficient n, which is called the roughness factor; seepage has its permeability coefficient k; and so on. All of these coefficients enable us to determine the rate of conversion of hydraulic energy into heat energy. This is the central question of hydraulics, but also it is the least understood one. Because of the complexity of the physical mechanism involved, we may not necessarily assume that the influence of various parameters on the rate of conversion is in linear relationship with certain major variables. Indeed, since in many instances the relationships cannot be represented by a proportionality constant, these factors are themselves dependent on other variables. This gives credence to the remark made by Theodore von Kármán, one of the greatest aerodynamics scientists of our age, who referred to hydraulics as "the science of variable constants." Even today, because of the complexity in most hydraulic applications, the influence of various parameters on the rate of conversion of hydraulic energy into heat energy is taken into account by empirical coefficients.

Many generations of hydraulicians strived to find better means to compute the rate of this energy transfer. More mathematical methods and less empirical ones would give better values than the empirical coefficients, which in many instances are only slightly better than educated guesses.

A major scientific breakthrough came early in our century when Ludwig Prandtl developed his *boundary layer theory*. The essence of the boundary layer theory is that it separates the main body of the flow, where surface effects are not felt, from the layers near the solid walls. As the fluid sticks to the wall surfaces the velocity within the boundary layer increases from zero at the wall to a value at which the effects of the surface can not be felt. The boundary layer theory allows engineers to write their theoretical formulas for fluid flow in a manner better representing the physical conditions of nature.

There are various formulas derived for generally the simplest cases of physical boundaries on the basis of the classical boundary layer theory. These give a better description for the velocity distribution curve than previously available. Many of these equations still contain some empirical coefficients because they were adjusted to better fit the experimental results available. But there are relatively few formulas that were derived for hydraulic applications on the basis of the boundary layer theory. The reason for this is the mathematical complexity of the theoretical formulas coupled with the physical complexity of most hydraulic problems.

One of the intrinsic laws of energy transfer is the *law of optimum energy*. In popular terms it states that "the water seeks the path of least resistance." Scientifically this means that of all possible flow distributions the one and only one that may occur in nature is the one requiring the least amount of energy per mass of water, that is, the one delivering the most fluid under the given conditions of energy availability. This law controls flows in pipe networks, open channels, and seepage, to mention only a few of the important situations. Solutions obtained by statistical methods using the so-called *probabilistic approach* indicate that the correct answer that satisfied this minimum energy theorem corresponds to the mean value of all possible answers. Most common methods of hydraulic analysis satisfy intrinsically this basic law. Where a minimization theorem is to be specifically stated, the computations are complex beyond the scope of our study.

Problems

4.1 A storage reservoir is filled through a pipe delivering 0.5 m³/s. The reservoir is circular in shape, with a diameter of 16 m. The height of the reservoir is 6 m. Determine the time rate of rise of the water surface and the time it will take to fill it. (*Ans.* 0.15 m/min, 40.2 min)

4.2 The rate of inflow into a river section is one-half of the outflow. The channel is rectangular in shape, 30 m wide. The length of the section considered is 1400 m. How long will it take to reduce the water level by 3 m if the outflow is 26 m³/s? (*Ans.* 2 hrs 42 min)

4.3 Water is to be raised by a pump against a total elevation of 60 m. How much work will be done if 5 m³/s is delivered for two hours? (*Ans.* 21.2 GN-m)

4.4 A 25 cm diameter 2 km long pipe line carries a discharge of 0.1 m³/s. The two ends of the line are at equal elevations. At the downstream end the total hydraulic energy is equal to 24 m above the pipe. If the slope of the energy grade line is 0.088 m/m, determine the pressure energy at the inlet point. (*Ans.* 200 m of water)

4.5 The bottom of a rectangular flume is 25 m above reference level. The depth of the water in the flume is 2 m. The average velocity is 0.5 m/s. Compute the total hydraulic energy at the surface of the water and at the bottom of the flume. (*Ans.* 27 m)

4.6 A horizontal pipe line is composed of a 12 cm diameter 500 m long pipe connected to a 25 cm diameter 700 m long pipe. The discharge in the pipe line is 0.01 m³/s. If the pipe discharges into the atmosphere and the rates of energy loss are 0.01 m/m and 0.002 m/m, respectively, determine the required total hydraulic energy at the inlet (neglect local losses). (*Ans.* 6.6 m)

4.7 The discharge in a 0.5 m diameter pipe is 3 m³/s of water. Determine the momentum of the flow. (*Ans.* 45.9 kN)

4.8 A 50 l/s discharge is flowing in a 25 cm diameter, 180 degree pipe elbow. Determine the magnitude and direction of the impulse force acting on the fixture. (*Ans.* 102 N)

4.9 A constricting 45 degree pipe bend is 10 cm in diameter at the inlet and 6 cm at the outlet. For a discharge of two liters per second, determine the change of momentum through the bend. (*Ans.* 1.18 N)

Chapter 5
Flow Measurements

5.1 Introduction

Whether in the field or laboratory, discharge and other measurements provide the fundamental data on which hydraulic analysis and design is based. Measuring discharges and other flow parameters is an essential part of the analysis and operational control of all hydraulic systems. The rate of use of water by municipal customers, industrial processes, and agricultural users is usually metered. The capacity of water courses and hydraulic structures must be determined. Hydrologic study of creeks and rivers is based on the statistical analysis of a long sequence of data obtained by repeated or continuous measurements. For these various applications a multitude of methods and devices were developed over the years. All of these are based on the fundamental physical laws of fluid mechanics discussed in Chapter 4.

Basically, flow determinations may be made directly and indirectly. *Indirect* determination of the flow involves defining or establishing known flow conditions and measuring one or more parameters, such as pressure or its variation, kinetic energy, and water surface elevations. The measured parameters along with the hydraulic formulas applicable define the flow rate.

5.2 Pressure Measurements

The *measurement of pressures* and pressure differences is performed by manometers, differential manometers, or Bourdon gages. The simplest method to measure pressure in a closed pipe is by the use of a transparent vertical or slanted standpipe, called *piezometer tube,* connected to a tap on the pipe in which the pressure is to be measured. By Equation 3.1 the pressure energy in the water will force the water level up to a height y, at which the static pressure in the piezometric tube equals that in the pipe itself, as shown in Figure 5.1. Piezometric tubes are used for some laboratory work,

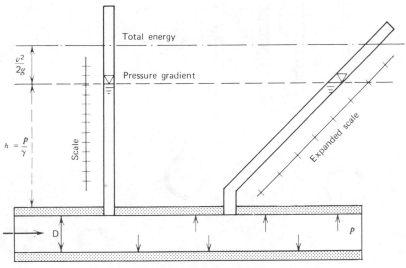

Figure 5.1 Vertical and slanted piezometric tubes.

but for practical applications they are not feasible. Since the height required is proportional to the unit weight of the fluid in the piezometer, large pressures could be measured more conveniently by using mercury that is 13.5 times heavier than water. To prevent the mercury from flowing out into the pipe the piezometer is bent into a U-shaped traplike loop. The resulting device is called a *manometer* and is shown in Figure 5.2.

The simple manometer shown in Figure 5.2a allows the determination of the pressure in the pipe at point m by writing

$$p_m = \gamma_{Hg}z - \gamma h \tag{5.1}$$

in which γ_{Hg} is the specific weight of mercury and γ is that of water. For a mercury manometer, therefore, Equation 5.1 takes the form of

$$p_m = \gamma(13.5z - h) \tag{5.2}$$

In this case we assumed that the pressure at point b is atmospheric; hence, p_m is the pressure above atmospheric pressure. If the other side of the manometer is connected to another pressure tap, as shown in Figure 5.2b, we speak of a differential manometer. A *differential manometer* is used to measure differences in pressure. Because of differences in pressure at the points m and n, there exists a difference in level, z, between the two mercury columns in the U-tube. By starting at m where the pressure is p_m, and noting the changes in

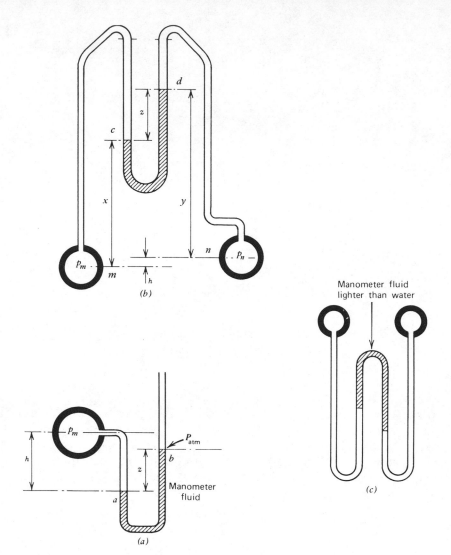

Figure 5.2 Manometers: (a) simple manometer; (b) differential manometer; (c) inverted manometer.

pressure as we pass to points c, d, and n, the following expression for p_n is obtained:

$$p_m - \gamma x - \gamma_{Hg} z + \gamma y = p_n$$

or

$$p_m - p_n = \Delta p = \gamma(x - y) + \gamma_{Hg} z$$

Referring to Figure 5.2b, one may see that

$$x - y = -h - Z$$

and by substitution

$$\Delta p = \gamma_{Hg} z - \gamma z - \gamma h$$

$$= \gamma[z(13.5 - 1) - h] \qquad (5.3)$$

In case the two pressure taps are at the same level, then $h = 0$ and

$$\Delta p = 12.5 \gamma z \qquad (5.4)$$

in case mercury is used as manometer fluid. For other manometer indicating fluids, Equation 5.3 must be rewritten accordingly.

When the pressure difference between two measured points is very small, the elevation difference in adjacent piezometer pipes is difficult to determine. It is convenient in such cases to magnify the difference by using a manometer fluid that is lighter than water. Various manometer oils are available commercially having specific weights about 80 percent that of water. Differential manometers designed to work these lighter than water manometer fluids are of the inverted kind, as shown in Figure 5.2c. There is a variety of manometer fluids used within the range of mercury and light oils. Usually these are color-coded such that their specific gravity may be ascertained beyond doubt. In the use of manometers care must be taken to expell all air bubbles from all connecting tubes, otherwise the readings obtained will have little value. For correct measurements the pressure taps in the pipes must be free of burrs and the entrance holes should be on a surface parallel to the velocity in the pipe.

Measurement of high pressure may also be made by ordinary steam gages invented by Bourdon. *Bourdon gages* consist of a coiled tube having its inner end closed and connected by a simple rack and pinion to a hand that is free to rotate over a graduated dial as shown in Figure 5.3. The pressure in the coiled tube tends to uncoil the tube, which results in a movement of the hand. The dial is calibrated by applying a known pressure to the tube and marking the position of the hand on the dial.

Modern technology developed a variety of pressure-sensing devices to be used in fluids. The most common types measure the deflection of a circular

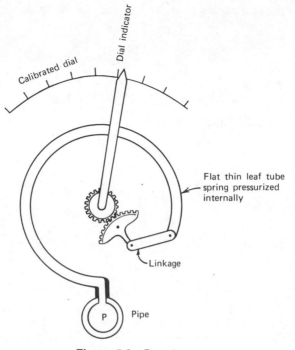

Figure 5.3   Bourdon gage.

elastic membrane under pressure. The deflection is transmitted to a gage or meter either mechanically, magnetically, or electrically.

Electrical linkage between the deflecting elastic membrane and a calibrated indicating meter or a recorder is provided by *pressure transducers*. Pressure transducers are usually built such that an electric strain gage of variable resistance is bonded to the membrane. The strain measured by the gage results from differential fluid pressure on the membrane. Hence the change in measured voltage flowing through the gage is proportional to the pressure of the fluid on the diaphragm. By simply changing the diaphragm the pressure range to be measured can be changed.

The advantages of electric pressure transducers in practice are based on the facts that such signals can be amplified to attain greater precision, and that they can be continuously recorded.

Example 5.1

A mercury-filled differential manometer between two adjacent pipes indicates $z = 5$ cm (1.97 in.) pressure. The pipes are carrying water and they are at the same elevation. What is the pressure difference between the pipes?

Solution

Substituting the specific weight of water into Equation 5.4 we have

$$\Delta p = 12.5(9810)0.05 = 6.131 \text{ kN/m}^2 \ (128.7 \text{ lb/ft}^2)$$

Example 5.2

A simple manometer, like the one shown in Figure 5.2a, is located such that the mercury level is 30 cm (11.8 in.) below the tap when the instrument indicates no pressure. Determine the pressure in the pipe if the difference in the levels of mercury is 4.5 cm (1.77 in.).

Solution

Equation 5.2 will apply with

$$h = 30 \text{ cm} + z/2 = 30 + 2.25 \text{ cm} = 0.325 \text{ m}$$

Substituting, we get

$$p_m = 9810(13.5 \ (0.045) - 0.325) = 2.77 \text{ kN/m}^2 \ (58.2 \text{ lb/ft}^2)$$

5.3 Velocity Measurements

Velocity measurements in pipes, as well as open channels, may be made by measuring the pressure corresponding to the kinetic energy of the flow. The simplest method is to measure the pressure in an open-ended tube bent in such a way that the end is aligned opposite the velocity vector measured. Such tubes are called *Pitot tubes*. The kinetic energy at the center of a Pitot tube converts into pressure energy as the flow of the fluid is halted at that point, as shown in Figure 5.4a. In pipes the pressure differential h between a Pitot tube and a nearby static tube provides the means to compute the velocity at the point measured such that

$$v = \sqrt{2g \ \Delta h}$$

in which $\Delta h = \Delta p/\gamma$. Hence

$$v = \sqrt{\frac{2g}{\gamma} \Delta p} \qquad (5.5)$$

By moving the Pitot tube across the diameter of the pipe the velocity distribution may be obtained from which the discharge can be calculated. Fixed Pitot tubes may be calibrated for a range of discharge to allow the discharge determination directly by multiplying the measured pressure differential by a constant.

The combination of static and Pitot tubes resulted in the *Prandtl tube* shown in Figure 5.4*b*. Prandtl tubes are available commercially in various sizes, some as small as 1.6 mm diameter.

Further improvement of this type of measurement is represented by the *Annubar*® shown in Figure 5.4*c*. Kinetic energy measured in the annubar is a composite of the velocities in several annuli of the pipe cross section. These measuring devices are calibrated directly in the flow by the use of a calibration curve.

When impulse-momentum principles are utilized, velocity measurements in pipes may also be made by *propeller meters,* or, more precisely, propeller-type current meters. Although most of these meters are used in rivers and open channels, they also are used for velocity measurements in hydraulic power stations, in which case several propeller meters are attached to a crossbar located in the conduit leading to the turbine.

The commonly used velocity measuring device in rivers is the *propeller meter* shown in Figure 5.5*a*. There are various propeller meters available commercially. A typical propeller meter consists of a propeller runner, a revolution counter, a shaft, and a tail piece. The propeller is usually seven to twelve centimeters in diameter. The pitch of the propeller is dependent on the average velocity for which the meter is designed. Normal propellers are designed to operate within the velocity range of 0.03 to 10 m/s. Care is taken to develop propellers such that they react to velocity components in their axial direction only. This enables the operator to measure the true velocity passing through the river cross section perpendicularly when the meter is held by a rigid rod. The counter gives an electric signal audible through an ear phone after each revolution or after every 5, 10, or 20 revolutions. The tail piece acts as a rudder, holding the meter in the direction of the flow, and is used only if the current is known to be perpendicular to the cross section measured.

The *Price current meter* used in the American practice is manufactured by Teledyne–Gurley Company and, in contrast to propeller meters, uses a cup-type anemometer rotating around a vertical shaft, as shown in Figure

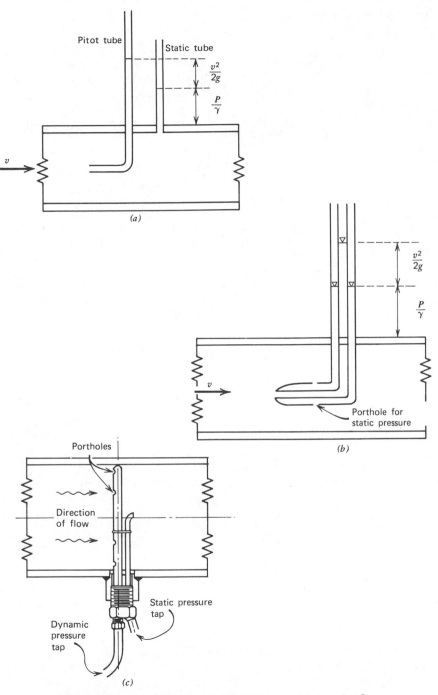

Figure 5.4 (a) Pitot tube; (b) Prandtl tube; (c) Annubar®.

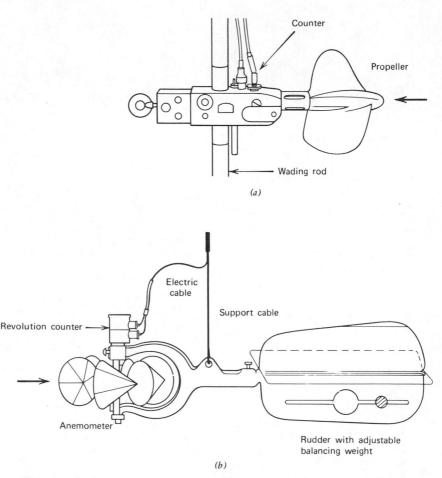

Figure 5.5 Current meters: (a) propeller meter; (b) Price current meter.

5.5*b*. It is manufactured in various sizes for different applications and may be used either on a wading rod or on a cable with a weight attached. The direct reading current meter includes a six-pole permanent magnet and a reed switch sending electric signals to a direct readout unit on the surface, which shows the velocity of the water on a dial indicator.

All current meters, whether of propeller or anemometer configuration, operate on the impulse-momentum theory. To use them one must have a calibration table. Calibration of these instruments is done in towing tanks where the instruments are attached to a carriage that is pulled over the tank at various velocities during which the rates of revolution of the meters are recorded. Under careful maintenance and proper operation the attainable precision with propeller-type current meters ranges between 5 and 10 percent, depending on the rate of pulsation in the current. The larger the pulsation in the stream, the greater the possible error.

Simple but approximate information on the velocity in an irrigation or drainage canal may be obtained by *surface floats*. For streams of medium size sometimes 15 to 25 floats are used simultaneously. These are equally distributed along the starter channel cross section. At a distance away from the starter section an arrival of the floats are recorded. The distance between the two sections should provide a travel time of several minutes, unless the velocity of the stream is rapid and the channel is meandering. Since the float velocity is representative of the surface velocity of the stream, the measured velocities should be reduced by a factor ranging from 0.75 to 0.9, the larger value being applied to deeper and faster (more than 2.0 m/s) streams. Somewhat more precise results may be obtained by *adjustable submerging floats* that may be made by thin aluminum tubes closed at both ends and filled with weight at one end to the degree that they float. Such floats are built such that they submerge to a depth of about 25–40 cm above the bottom and stick out over the surface by about 5–10 cm. The observed velocity of a submerging float, v_f, allows the determination of the average stream velocity v_{ave}, along its path by the experimental formula

$$v_{ave} = v_f (0.9 - 0.116 \sqrt{1 - h_f/y}) \tag{5.6}$$

where h_f is the length of the float and y is the depth of the stream.

Float-type measurements are sometimes used for large rivers also. This procedure may give valuable information on stream line patterns at a cost that is considerably less than procedures using current meters. For large dam projects in Europe and India floating torches were used at night and their paths were recorded by continuous photographing from several nearby hills. The results were reduced by using techniques known from descriptive geometry.

Example 5.3

The manometer connecting the static and Pitot tubes of a velocity probe indicates 5.8 kN/m² (121.8 lb/ft²) pressure. How much is the velocity measured?

Solution

Equation 5.5 will apply with a specific weight of 9.81 kN/m³ and the pressure differential of 5.8 kN/m². Substituting we obtain

$$v = \sqrt{\frac{2(9.81)}{9.81}} (5.8) = 3.4 \text{ m/s (11.2 ft/sec)}$$

5.4 Discharge in Pipes

Direct determination of the discharge could be made by volumetric or weight measurements. The simplest way of measuring a discharge is to allow the flow into a container of known volume and then measure the time required to fill it. In laboratories containers placed on scales allow the weighing of the water entering the container in a measured time interval. The measured weight, divided by the unit weight of water and by the measured time interval, gives the flow rate in volume per unit time. Municipal water meters measure the volume of water used over a period by direct measurement. These are called positive displacement meters because they measure the volume of water that repeatedly fills a given container. Usually the number of fillings is counted by a register or counter and the total quantity is obtained as the product of the volume of a chamber and the number of fillings. The registers are geared such that the counters are showing the total volume in gallons, cubic feet, or liters. Hence the displacement meters may be looked upon as fluid motors operating at a high volumetric efficiency under a very light load. These types of water meters, which include the oscillating or rotary piston meters, sliding or rotating vane meters, nutating disc meters, gear or lobed impeller meters, and other devices, are named according to their mechanical design. In addition to their importance in the determination of customers' water use, which is of little interest to the hydraulic designer, these meters that register positive displacement were introduced recently in a broad range of hydraulic discharge measurements. By replacing head loss measuring and registering devices in orifice- or flume-type hydraulic measuring structures, they serve as proportional secondary circuit meters.

Indirect flow measurements in pipelines are made, in practice, by a great variety of commercially available measuring devices. All of them use the principle of continuity and one other equation that defines the motion, usually the energy equation. Orifices, measuring nozzles and venturi tubes, Pitot tubes, Prandtl tubes, and Annubars® operate on this principle. Other devices, such as the propeller meters and elbow meters, use the momentum equation.

In a pipe elbow the direction of the velocity changes and, by virtue of the momentum theorem, an impulse force is generated along the outer curvature of the bend. By measuring the difference in pressure between the inner and outer curvature of the pipe bend, as shown in Figure 5.6, the effect of the impulse force is registered. By proper calibration the discharge in a pipe elbow of a given diameter may be determined. An approximate discharge formula for 90 degree pipe elbows, for which the radius of curvature of the pipe R is larger than twice the pipe diameter, may be expressed as

$$Q = \frac{\pi D^2}{4} \sqrt{\frac{R}{2D}} \sqrt{2g} \sqrt{\frac{\Delta p}{\gamma}} = 0.785 \frac{D^{1.5}R^{0.5}}{\rho^{0.5}} \sqrt{\Delta p} \qquad (5.7)$$

where Δp represents the pressure difference between the inside and outside pressure taps located along the 45° diagonal of the elbow. For continuous discharge measurements pipe elbows may be equipped with a standard positive displacement water meter known from municipal metering practice. In such arrangements the difference in pressure between the taps will force some water to flow across the meter. This discharge, registered by the meter, will be proportional to the pressure difference, which, in turn, varies directly with the discharge in the pipe. The system, once calibrated, is a useful way to determine the output of pumping installations.

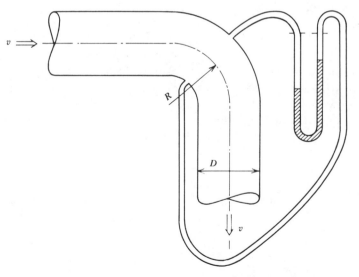

Figure 5.6 Elbow meter equipped with manometer.

The simplest method of measuring discharge in a pipe is offered by a *circular orifice*. Its arrangement is shown in Figure 5.7. By virtue of the equation of continuity one may observe that

$$Q = (V \cdot A)_{\text{pipe}} = (v \cdot a)_{\text{orifice}} \tag{5.8}$$

If we introduce the energy equation with the assumption that the pipe is horizontal, the Bernoulli equation takes this form:

$$\frac{V^2_p}{2g} + \frac{p_p}{\gamma} = \frac{v^2_0}{2g} + \frac{p_0}{\gamma} + h_2 \tag{5.9}$$

where h_2 is an energy loss occurring at the orifice edges and in the turbulent exit zone behind the orifice plate. For orifices and similar flow meters the energy loss is expressed as a coefficient c, which is experimentally recorded at the time of calibration. Combining Equations 5.8 and 5.9 and rearranging the terms, one may write the ideal average velocity through an orifice as

$$v = \sqrt{\frac{(2g/\gamma)(p_p - p_o)}{1-[c(a/A)]^2}} \tag{5.10}$$

and the discharge Q as

$$Q = \frac{ca}{\sqrt{1 - c^2(d/D)^4}} \sqrt{\frac{2g}{\gamma} \Delta p} \tag{5.11}$$

where d and D are the diameters of the orifice and the pipe, respectively, and

$$\Delta p = p_p - p_0 \tag{5.12}$$

Experimental studies indicated that the actual value of the c coefficient depends on the location where p_0 is measured. In addition it also depends on the configuration of the edge of the orifice, on the ratio d/D, and on the velocity. A good approximate formula for orifices in practical applications may be written in the simple form of

$$Q = ca \sqrt{\frac{2g}{\gamma} \Delta p} \tag{5.13}$$

The value of c is to be determined by calibration. Commercially available orifice meters are supplied with a calibration chart. The discharge coefficient is usually within the range of 0.6 to 0.7. For water the discharge equation of a given orifice can be represented by the simple formula of

$$Q = K \Delta p^{1/2}$$

where K is a parameter lumping all variables that are fixed for the orifice in question.

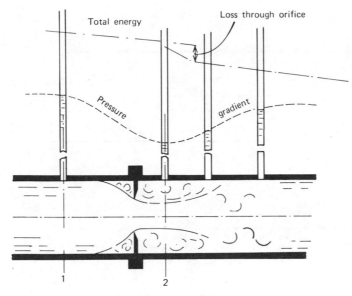

Figure 5.7 Flow through a circular orifice meter.

To reduce the losses occurring in the constricted flow and to extend the range of the constant values of the coefficient, orifices are sometimes substituted by nozzles or venturi tubes as shown in Figure 5.8.

The derivation of the discharge formula for *flow nozzles* or *venturi tubes* is identical to that for the orifice meter. Discharge coefficients of these devices range from 0.94 to 0.98.

In metering any fluid in a pipeline, it is most important that the flow approach the orifice, flow nozzle, venturi, or any other metering element in a normal state. The flow must not be influenced by swirls, cross currents, eddies, or other disturbances that create helical paths of flow. Disturbances after the flow-measuring element may introduce errors into the static pressure measurements at that point. To produce uniform conditions it is recommended that a length of straight pipe at least six to ten times the pipe diameter be in front of the measuring devices. If this is not possible in some installations, then guide vanes are to be installed in the pipe to straighten the flow. After the metering device, a straight length of three to five times the pipe diameter is desirable.

Another class of instruments measuring discharge in piping systems is the *variable area meter*. These include the tapered tube and float meter, the multiple-holed cylinder and piston, and the slotted cylinder and piston types.

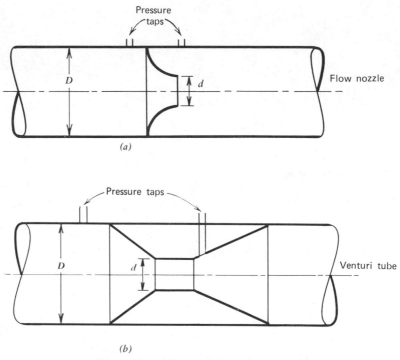

Figure 5.8 Flow nozzle and venturi tube.

In all variable area meters the energy causing the flow through the meter is substantially constant for all discharges and the discharge is directly proportional to the metering area. In the *tapered tube and float meter,* shown in Figure 5.9, the area variation caused by the upward flowing fluid produces a rise or fall of the float in the tapered tube. The float is designed so that the impulse force and viscous drag force raising it, less the force of buoyancy, are constant for all positions of the float in the tapered tube. These instruments are referred to as rotameters. In the Fisher–Porter laboratory rotameter the tapered tube is made of glass to make the position of the float visible. The discharge versus float elevation is determined by calibration. Since the viscous drag force lifting the float is dependent on the viscosity of the fluid, which in turn is dependent on the temperature, for different fluids or calibration fluid at different temperature the actual readings of the meter must be reduced to calibration data. The actual discharge q_a of a fluid with actual unit weight corresponds to calibration discharge q_c as

$$q_a = q_c \sqrt{\frac{\gamma_c}{\gamma_a}} \tag{5.14}$$

where γ_c is the unit weight of the fluid for which the meter was calibrated. The *slotted cylinder-type variable area meter* utilizes a cylinder with a slot on its side through which the flow exits the meter. The varying rates of flow will cause a piston in the cylinder to rise or fall. The position of the piston is seen through a sight tube in front of a calibrated scale.

Faraday's principle of magnetic induction is used directly for discharge measurement in *magnetic flow meters*. Other new pipe flow measuring devices include the *electrostatic flow meter*, based on two dissimilar metal pipe sections connected by a nonconducting pipe section. The voltage change between the two metals relates to the flow rate of the fluid. Also under development is the *thermal wave flow meter*, which correlates the time of flow of a periodic thermal wave between an electric heater element and a sensor.

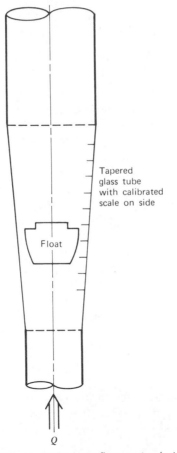

Tapered glass tube with calibrated scale on side

Float

Q

Figure 5.9 Variable area flow meter (rotameter).

Example 5.4

The pressure difference across a 6-cm (2.36-in.) diameter orifice is 12 kN/m² (252 lb/ft²). The discharge coefficient of the orifice was calibrated to be 0.65. Determine the discharge of the water.

Solution

The cross-sectional area of the orifice is

$$a = \pi d^2/4 = 0.0028 \text{ m}^2$$

To determine the discharge Equation 5.13 will be used. Substituting all variables we have

$$Q = 0.65(0.0028) \sqrt{\frac{2(9.81)}{9.81}} \; 12 = 0.0089 \text{ m}^3/\text{s (141 gal/min)}$$

5.5   Discharge in Open Channels

Direct determination of river discharge is usually done by concurrent measurements of depths and velocity distributions at a suitable cross section of the stream. The best location to perform such measurements along a stream is where the direction of the velocity is substantially perpendicular to the section, cross currents are minimal, and the cross section is reasonably uniform and free from vegetable growth. Bridges are ideal locations for such measurements, otherwise the use of guy-cables and boats are required.

In shallow rivers where the depth does not exceed three to four meters and the velocity is less than 0.5 m/s, graduated *sounding or wading rods* are used for measuring the depth. Deeper channels are measured by *cables with weights* at their end. The size of these weights depends on the maximum velocity expected in the stream. For velocities ranging up to 0.5 m/s, a 5 kg weight is sufficient; for 2 m/s, a 50 kg weight is necessary. For high velocities up to 4 m/s, 100 kg weights are recommended. To handle these large weights the measurement is performed by utilizing motorized winches.

More precise measurements of depth may be performed by *ultrasonic devices*. These measure the time required for a sound pulse emitted by the instrument to bounce back from the bottom to a receiver, similar to the way sonar operates.

For a known channel cross section the velocity is measured along a number of verticals. For medium size rivers at least 10 to 15 verticals are measured distributed evenly across the channel. At each vertical velocities are measured at several depths. For shallow depths one point measurement at $0.6y$ is sufficient. For depths up to 0.6 m, two points measured at $0.2y$ and $0.8y$ are generally sufficient. In these cases the average velocity is determined by

$$v_{ave} = \frac{v_{0.2} + v_{0.8}}{2} \tag{5.15}$$

In three-point measurements at depths of $0.15y$, $0.5y$, and $0.85y$, the average velocity is still the arithmetic mean of the three measured values. For uneven velocity distribution along a deep vertical the measurements are taken at five points. In addition to the measured velocity values near the surface, v_s, and near the bottom, v_b, measurements are taken at $0.2y$, $0.6y$, and $0.8y$. The average velocity in this case is computed by

$$v_{ave} = \frac{v_s + 3v_{0.2} + 2v_{0.6} + 3v_{0.8} + v_b}{10} \tag{5.16}$$

Results of discharge measurements in a river can be graphically represented in a plot similar to the one in Figure 5.10. The discharge Q may be determined from the data by computing the discharge q in each strip between two adjacent verticals. In this case the discharge of a strip between two verticals is

$$q = \left(\frac{v_1 + v_2}{2} \right)\left(\frac{y_1 + y_2}{2} \right) \cdot b \tag{5.17}$$

where b is the width between the adjacent verticals and subscripts 1 and 2 refer to the measurements taken in the verticals. The total discharge is then the sum of that at all strips.

A more precise result may be obtained by graphical solution. In this, the velocities are plotted on the cross section where they were measured and points of equal velocity are connected by contour lines. The areas of known velocities may be measured by a planimeter. The sum of all areas multiplied by their respective velocities give the total discharge of the stream.

Performing the discharge measurement of a river by the direct method described above is a time-consuming process. As the flow in a river is rarely constant, either rising or falling, the measured velocities must be adjusted to an average value, valid for the duration of the measurement. This adjustment must be carried out if the average depth of the stream changes by more than 1 percent during the measuring time. To establish the change of stage it is recommended that the water level be marked by a stake at the shore before and after the measurement, and that the change of average water level during this time be recorded.

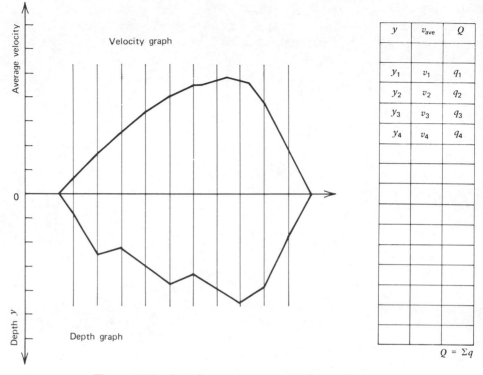

Figure 5.10 Data from measurement of river discharge.

Shallow rocky rivers, creeks, or small open channels cannot be measured in the same manner as rivers because of their small size. In the absence of hydraulic structures that may be utilized as points of measurement, either portable weirs or flumes are utilized, or the discharge is determined by chemical means.

Weirs or flumes are often constructed at the site. With proper calibration such structures may give reliable information on the flow rate, but their shortcomings are significant. If the soil in which they are placed in the channel is pervious, the prevention of seepage around them is almost impossible. The requirement of free overfall makes it necessary to dam up the upstream water, resulting in considerable pressure on the upstream side of the structure, which will require significant structural supports. In cases when these portable weirs are applicable, their discharges are determined by the weir discharge formula. The discharge coefficient is generally determined by calibration.

For small discharges in laboratories or man-made canals portable weir plates with 90 degree or 60 degree V-notch weirs—the so-called *Thomson weirs*—are often used. The discharge formulas for these are in the form of

$$q = mg^{0.5} h^{2.5} \qquad (5.18)$$

in which the discharge is in m^3/s and h, the depth of water in the weir, is in meters. The factor m depends on the shape of the weir. For 90 degree weirs it is 1.17; for 60 degree weirs it is 0.67.

The above equations are valid for smooth, knife-edged weir plates. Experience shows that the V-notch weirs are very sensitive to any change in the roughness of the weir plate; hence the equations can be assumed to give only approximate values under practical conditions.

For larger discharges the trapezoidal *Cipoletti weir* is used with a base width of b and steep side slopes with a 4 to 1 rise. Figure 5.11 allows the solution of Cipoletti's discharge formula to be done in a graphical manner.

Chemical discharge measurements are most valuable for small streams where a large part of the flow passes through the rocks and gravel of the stream bottom. The common principle of the chemical discharge measuring techniques is based on the continuity equation. It involves the introduction of a known amount of tracer chemical into the water, which, after traveling in the stream over a certain length, will completely diffuse into the water. The amount in which the tracer chemical is found in the water after complete mixing is proportional to the discharge of the stream. Hence, by determining the degree of saturation of tracer in the water, its discharge may be computed. Stream discharges can also be measured by injecting a fluorescent dye (Rhodomine) and measuring its concentration at a point downstream.

Example 5.5

A 90 degree Thomson weir discharges water such that the depth in the weir is 70 cm (27.6 in.). Determine the discharge.

Solution

Equation 5.18 applies with an m value of 1.17. The discharge then will be

$$Q = 1.17(9.81)^{0.5}(0.70)^{2.5} = 1.5 \ m^3/s \ (53 \ ft^3/sec)$$

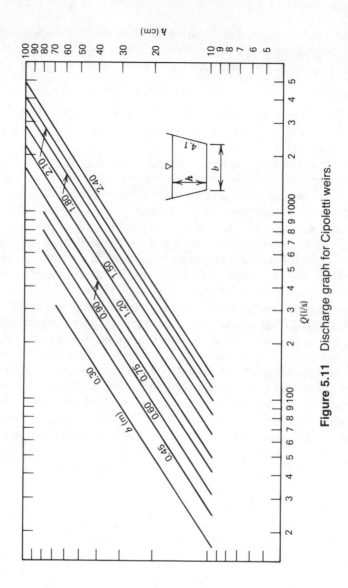

Figure 5.11 Discharge graph for Cipoletti weirs.

Example 5.6

A Cipoletti weir with a 1.5 m (4.92 ft) base carries $h = 0.5$ m (1.64 ft) water. How much is the discharge?

Solution

The graph shown in Figure 5.11 may be utilized to solve this problem, with $b = 1.5$ m and $h = 0.5 = 50$ cm. The result is 600 l/s, or 0.6 m³/s (21.2 ft³/sec).

Problems

5.1 Explain the difference between direct and indirect flow determinations.

5.2 Formulate an equation, similar to Equation 5.4, for a 45-degree slanted mercury manometer.

5.3 Two pipes carrying water are 68 cm apart vertically. A differential manometer between them indicates 17 cm of mercury as pressure differential. What is the pressure difference if the mercury is displaced toward the lower pipe? (*Ans.* 27.5 kN/m²)

5.4 A Prandtl tube indicates a 45 cm mercury pressure differential. What is the velocity measured? (*Ans.* 10.9 m/s)

5.5 A stream is 3 m deep. A surface float submerges 40 cm deep. The measure time of passage over a 300 m distance along the stream was 10 minutes. Determine the average velocity of the stream. (*Ans.* 0.43 m/s)

5.6 An elbow meter is measuring water. The pipe diameter is 20 cm and the radius of curvature of the pipe is 40 cm. Determine the expected difference in pressure for a discharge of 0.08 m³/s. (*Ans.* 331 N/m²)

5.7 A flow nozzle is 30 cm in diameter, installed in a 72 cm discharge pipe. The discharge coefficient is 0.92. What is the discharge if the measured pressure differential is 15 kN/m²? (*Ans.* 0.356 m³/s)

5.8 A 60 degree Thomson weir is submerged 40 cm above its notch. Determine the discharge flowing. (*Ans.* 0.212 m³/s)

5.9 A Cipoletti weir is 60 cm wide. The depth of flow is 70 cm. What is the discharge? (*Ans.* 650 l/s)

Chapter 6
Pipe Flow

6.1 Basic Concepts

In hydraulics any closed conduit that carries fluid under pressure is considered to be a pipe regardless of its cross-sectional shape. Usually a pipe has a circular cross section. Closed conduits that are not flowing full are treated as open channels. Sewers, culverts, and drainage pipes rarely flow full and under pressure; therefore pipe flow computations normally do not apply to them.

Pipes are manufactured in certain selected diameters. In actual applications the nearest existing size is chosen in terms of the so-called *nominal diameter*. As shown in Table 6.1, the nominal diameter does not necessarily represent the actual inside diameter of the pipe. The nominal diameter is usually quoted in inches, even in countries where the metric tradition is old.

Pipe flow computations are generally aimed at determining the sum of the energy losses incurred while delivering fluids from one point to another at a specified pressure and in a specified quantity. Either from the potential energy of a sufficiently elevated reservoir or from the pressure energy delivered by a pump, the energy input must overcome the energy losses incurring. The factors affecting losses of energy in pipe flow are independent of the pressure. The most important parameter that influences these energy losses is the kinetic energy of the flow, $v^2/2g$. Other influencing parameters are geometrical ones, including the length L and the diameter D of the pipe, for the most part.

There are two kinds of energy losses in pipe flow: local losses and friction losses along the pipe. Both are caused by the viscous resistance of the fluid. Local losses, discussed in detail in Section 6.3, occur where there are sudden geometric changes in the conduit, such as elbows, valves, sudden changes in diameters, and the like. Friction losses along the conduit are caused by the roughness of the pipe wall and by shear between the fluid particles as they move down the pipe at different velocities. In both cases, the kinetic energy is the primary controlling variable.

Table 6.1
Nominal and Actual Inside Diameters of Selected Pipes

Nominal diameter (in.)	Actual inside diameter (cm)	
	Seamless steel, Schedule S	Cast iron, Class A
4	10.22	10.06
6	15.40	15.29
8	20.27	20.65
10	25.45	25.65
12	30.48	30.78
24	59.05	61.67

For all but very slow movement in very small diameter pipes, there is turbulent flow in pipes. In turbulent flow, fluid particles follow independent and very random paths. The concept of average velocity is only a statistical concept; on the average a given Q discharge passes through the cross-sectional area A of the pipe at a certain point. This turbulent movement of the fluid particles accounts for a large measure of the energy losses in pipe flow. The degree of turbulence grows with the increase of velocity. One measure of turbulence is a dimensionless term called the *Reynolds number*, **R,** which is defined as

$$R = \frac{V \cdot D}{\nu} = \frac{V \cdot D \cdot \rho}{\mu} \tag{6.1}$$

in which V is the average velocity of the flow, D is the diameter of the pipe, and ν is the kinematic viscosity of the fluid. For water at room temperature the kinematic viscosity may be taken to be 10^{-6} m²/s as a good approximation.

For flows when the Reynolds number is less than 2000 the turbulence is suppressed and the fluid particles move along in ordered parallel paths. Such flows are called laminar.

In pipe flow computations there are two fundamental equations utilized: the equation of continuity and the Bernoulli equation. The continuity equation allows the computation of velocities at any point of the pipe line where the diameter is given, as long as the discharge is known. In Figure 6.1, for example, the velocities in the successive pipe sections A, B, and C can be defined in relation to each other by writing

$$Q = \frac{\pi D^2_A}{4} v_A = \frac{\pi D^2_B}{4} v_B = \frac{\pi D^2_C}{4} v_C \tag{6.2}$$

or in the form of

$$v_B = v_A(D_A/D_B)^2 = v_C(D_C/D_B)^2 \qquad (6.3)$$

The latter equation makes the solution of Bernoulli's equations written for a sequence of interconnected pipes easier. Such a problem is depicted in Figure 6.1. Here the water flows by gravity from an upper reservoir through three pipes of varying diameters to a point (point e and 14 on the drawing) where it discharges freely into the atmosphere; the free jet will come to rest on the surface of a lower reservoir (point f and 17 on the drawing). The horizontal line connecting points 1, 6, and 15 represents the energy originally available at the upper reservoir. The line connecting points 1, 2, 4, 5, 7, 8, 10, 11, 13, 16, and 17 represents the *energy grade line*. Within the energy grade line sections 1–2, 4–5, 7–8, and 10–11 are local losses due to the inlet, the expansion and constriction, and the elbow in pipe C, respectively. The interconnected line marked by g, h, i, k, and e shown is the *hydraulic grade line*. The elevation

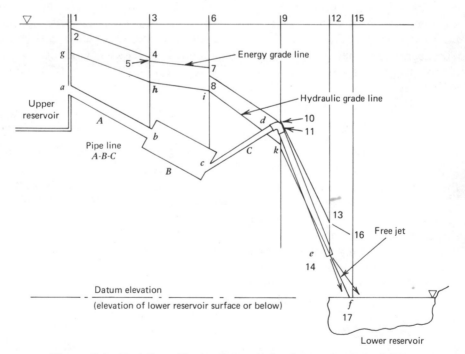

Figure 6.1 Energy and hydraulic grade line for a sequence of pipes.

differences shown between a–g, b–h, c–i, and k–d are pressure elevations at the locations a, b, c, and d. There is a notable situation at point d. Here the hydraulic grade line is below the pipe, at k, which indicates that the pressure in the pipe is less than atmospheric. The hydraulic grade line can always intersect the pipe elevation; however, the energy grade line must always be higher than the pipe itself. At the exit at point e one may note that there is atmospheric condition; therefore, the pressure at the end of the pipe is zero. As the water continues to flow in the form of a free jet, the energy grade line continues from point 13 to 16, indicating only a small loss due to friction with still air. At point f the jet comes to rest on the surface of the lower reservoir. The waves and turbulence caused by the impact of the jet will dissipate the energy of the flowing water. It must be mentioned that as long as the pipe cross section remains the same, the slope of the energy grade line is always parallel with the hydraulic grade line. Since there exists but one pressure at any point in the pipe, the hydraulic grade line is a continuous one. The energy grade line, on the other hand, may show sudden drops that represent local losses occurring. For most pipe flow problems involving different pipe sections and fixtures causing local losses the construction of drawings such as Figure 6.1 is always advantageous. It reduces the possibility of conceptual errors and omissions in the calculations.

Using the notations in Figure 6.1, the energy equation may be written for the first pipe (pipe A) as follows:

$$\frac{v^2_a}{2g} + \frac{p_a}{\gamma} + z_a = \frac{v^2_b}{2g} + \frac{p_b}{\gamma} + z_b + h_L \tag{6.4}$$

The terms in Equation 6.4 are, from left to right: the kinetic energy term for the first pipe, marked as the distance between points g and 2 (or points h and 4); the pressure energy at the beginning of the pipe, shown as the distance between a and g; the elevation of point a over the datum plane, z_a; the kinetic energy at the end of the pipe, same as at the beginning; the pressure energy at the end, marked as the distance between b and h, the elevation of point b over the datum plane, z_b. The last term, h_L, is the energy lost through friction over the pipe length L from points a to b. The determination of this last term will be the subject of Section 6.2 below.

Formulas similar to Equation 6.4 can be written for the other two pipes (B and C) of the pipeline problem shown in Figure 6.1. Adding formulas necessary to solve for local losses at points a, b, c, d, and e and introducing the equation of continuity shown in Equation 6.2 or 6.3, the total energy loss between the beginning and the end of the pipeline can be computed easily.

6.2 Frictional Losses in Pipes

The fundamental formula used in determining losses due to friction along pipes is the Darcy–Weisbach equation. It states that the energy loss h_L in a pipe is directly proportional to the length L and the kinetic energy, $v^2/2g$, present, and inversely proportional to the diameter of the pipe, D. The formula is written as

$$h_L = f \, \frac{L}{D} \, \frac{v^2}{2g} \tag{6.5}$$

in which the constant of proportionality, f, is called the *friction factor*. The friction factor itself is a rather complex function of the flow parameters, the kinematic viscosity of the fluid flowing, and the degree of roughness of the pipe wall. Since in the original scientific studies of this problem the pipe wall roughness was created by coating the pipe walls with sand of known diameter, the roughness of commercially available pipe materials was defined in terms of an *equivalent sand roughness*. Values for this equivalent sand roughness for selected pipe materials are listed in Table 6.2. By dividing the equivalent sand roughness with the inside diameter of the pipe, we obtain a dimensionless number called *relative roughness*.

The relationship between friction factor and other parameters of flow was subject to very extensive research in laboratories and the field for well over a century. Theoretical investigations of turbulence in pipes has contributed significantly to the clarification of this complex problem. By conveniently

Table 6.2
Equivalent Sand Roughness for Pipes

Commercial pipe surface, new	e (meter)	$\nu \cdot e$	$\dfrac{g \cdot e^3}{\nu^2}$
Glass, lucite, copper	smooth	—	—
Steel, wrought iron	0.5×10^{-4}	0.5×10^{-10}	1.2
Asphalted cast iron	1.2×10^{-4}	1.2×10^{-10}	17
Galvanized iron	1.5×10^{-4}	1.5×10^{-10}	33
Cast iron	2.5×10^{-4}	2.5×10^{-10}	153.3
Concrete pipe	$10^{-3} - 10^{-2}$	$10^{-9} - 10^{-8}$	$10^4 - 10^7$

$\nu = 10^{-6} \ \mathrm{m^2/s}$
$g = 9.81 \ \mathrm{m/s^2}$

plotting both theoretical and experimental results in a single graph, the Moody diagram, shown in Figure 6.2, permits the quick determination of the friction factor as long as the relative roughness of the pipe and the Reynolds number of the flow is known. The Darcy–Weisbach equation (Eq. 6.5) and the Moody diagram (Fig. 6.2) can be used directly for those problems where the energy loss, h_L, is sought, and the discharge as well as the pipe is defined. For such cases the relative roughness, e/D, and the Reynolds number from Equation 6.1 are computed first. By entering the computed values into Figure 6.2, the corresponding friction factor f is obtained. By substituting this result along with the other known flow parameters into Equation 6.5, the energy loss in the pipe can be determined directly.

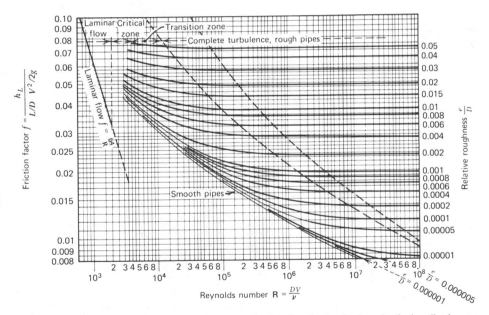

Figure 6.2 Moody's diagram to determine the friction factor f when both the discharge and the pipe diameter are known.

Example 6.1

A 65-m (213.2-ft) long 250 mm (9.84 in.) diameter galvanized iron pipe carries water at 0.04 m³/s (1.4 ft³/sec) discharge. Calculate the energy loss due to friction.

Solution

The equivalent sand roughness of galvanized iron, from Table 6.2, is $e = 1.5 \times 10^{-4}$ m. The relative roughness equals

$$e/D = \frac{1.5 \times 10^{-4}}{0.25} = 0.0006$$

The cross-sectional area of the pipe is

$$A = \frac{\pi D^2}{4} = 0.05 \text{ m}^2$$

The average velocity of the flow is

$$v = \frac{Q}{A} = \frac{0.04}{0.05} = 0.8 \text{ m/s}$$

The Reynolds number may be computed from Equation 6.1 using an approximate value for the kinematic viscosity for water at normal temperatures as 10^{-6} m²/s, resulting in

$$R = \frac{v \cdot D}{10^{-6}} = 0.8(0.25)10^6 = 2 \times 10^5$$

Entering Figure 6.2 at $e/D = 0.0006$ and $R = 2 \times 10^5$, one can find the corresponding friction factor to be

$$f = 0.02$$

With these the Darcy–Weisbach equation, Equation 6.5, may be written as

$$h_L = f \frac{L}{D} \frac{v^2}{2g} = 0.02 \frac{65}{0.25} \cdot \frac{0.8^2}{2(9.81)} = 0.169 \text{ m}$$

or approximately 17 cm (6.7 in.) Accordingly, the energy loss through the pipe is 0.17 m, and the slope of the energy grade line is

$$\frac{h_L}{L} = 0.17 / 65 = 0.026 \text{ m/m}$$

The gradient of the hydraulic grade line is the same. The vertical elevation difference between the energy grade line and the hydraulic grade line equals $v^2/2g$, which is 0.032 m (1.26 in.)

Example 6.2

A 300-m (984-ft) long 0.2 m (7.9 in.) diameter steel pipe connects two reservoirs. The upstream reservoir is located 200 m higher than the downstream one. How much energy is needed to be delivered by a pump in order to supply 0.05 m³/s (792 gpm) discharge (see Fig. E6.2)?

Solution

From Table 6.2 the equivalent sand roughness of steel pipe is 0.5×10^{-4} m. The relative roughness then is

$$e/D = 0.5 \times 10^{-4}/0.2 = 0.00025$$

The cross-sectional area of the 0.2 m diameter pipe is

$$A = \pi D^2/4 = \pi \, 0.2^2/4 = 0.031 \text{ m}^2$$

The average velocity of the flow is

$$v = Q/A = 0.05/0.031 = 1.61 \text{ m/s}$$

The kinetic energy of the flow is

$$v^2/2g = 1.61^2/19.62 = 0.132 \text{ m}$$

The Reynolds number is

$$R = vD/\nu = 1.61(0.2)10^6 = 3.22 \times 10^5$$

Entering Figure 6.2 with these results, the friction factor is found to be

$$f = 0.017$$

Substituting into the Darcy–Weisbach equation,

$$h_L = f(L/D)v^2/2g = 0.017(300/0.2)0.132 = 3.36 \text{ m}$$

In addition to overcoming the energy loss due to friction (local losses are neglected here), the pump must supply sufficient energy to raise the water to the upper reservoir, which is 200 meters higher. Furthermore, the pump must supply the kinetic energy for the motion. Therefore, the required pumping height is

$$H_{\text{pumping}} = h_L + (z_{\text{outlet}} - z_{\text{inlet}}) + v^2/2g$$

$$= 3.36 + 200 + 0.132 = 203.5 \text{ m}$$

In actuality the pump will have less than 100 percent efficiency and it will have to be considered also.

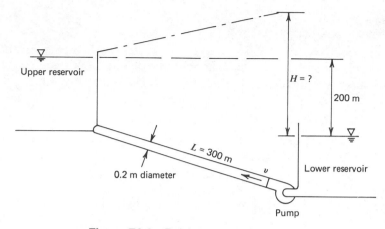

Figure E6.2 Raising water with a pump.

Moody's diagram and the Darcy–Weisbach equation do not allow for a direct solution when the pipe diameter is unknown. In such cases neither the relative roughness nor the Reynolds number is known at the beginning. Likewise, when the unknown is the discharge, the Reynolds number cannot be defined. To overcome computational difficulties, Equation 6.5 and Figure 6.2 were combined into Figure 6.3. The resulting graph is called the Asthana's diagram. When it is used to directly solve problems in which the discharge or the pipe diameter are unknown, recourse to the use of the Darcy–Weisbach formula is unnecessary. Table 6.2, along with the relative roughness values for commercial pipes, also includes the two functions $\nu \cdot e$ and $g \cdot e^3/\nu^2$, computed for water, that are used as multipliers along with Q and h_L/L on the coordinates of Figure 6.3. These will be found to simplify the use of Asthana's diagram when it is applied for flows of water.

Example 6.3

Two reservoirs whose surface elevation is different by 20 m (65.6 ft) are connected by a 800-m (2624-ft) long steel pipe. Select the diameter of the pipe such that it delivers a discharge of 0.5 m³/s (17.6 ft³/sec).

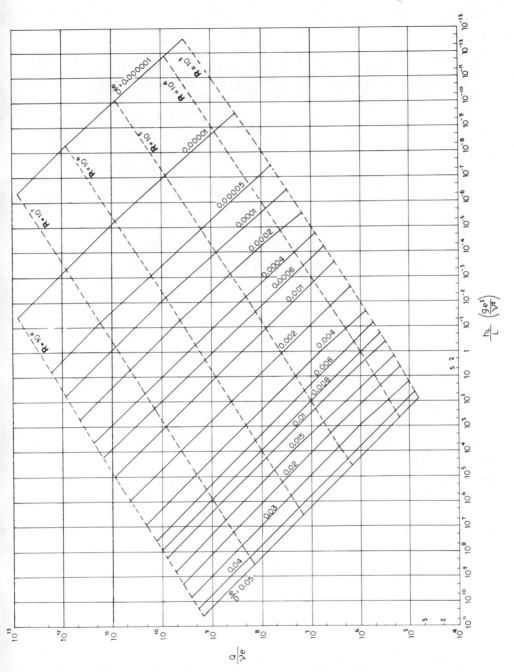

Figure 6.3 Asthana's diagram for pipe flow computation. (Courtesy of K. C. Asthana.)

103

Solution

From Table 6.2 the following data is found for steel pipes:

$$e = 0.5 \times 10^{-4} \text{ m}$$

$$e \cdot \nu = 0.5 \times 10^{-10} \text{ m}^3/\text{s}$$

$$g \cdot e^3/\nu^2 = 1.2$$

The elevation difference between the two reservoirs provides the energy needed for the motion ($\nu^2/2g$) and the energy to be lost by friction along the pipe. The kinetic energy term will be neglected here, as well as the local losses at the entrance and exit of the pipe. Hence

$$h_L = 20 \text{ m and } h_L/L = 20/800 = 0.025$$

The scale reading at the horizontal coordinate of Asthana's diagram is then

$$\frac{h_L}{L} (g \cdot e^3/\nu^2) = 0.025 (1.2) = 3 \times 10^{-2}$$

The reading on the vertical coordinate shall be

$$\frac{Q}{e \cdot \nu} = \frac{0.5}{0.5} 10^{10} = 10^{10}$$

Entering the diagram with these two readings one finds that

$$e/D = 0.00012 \simeq 10^{-4}$$

therefore $D \simeq e/10^{-4} \simeq 0.5$ m (1.64 ft) is the diameter of the pipe needed. Now one may check for the velocity and kinetic energy by computing

$$\nu = \frac{Q}{\pi D^2/4} = 0.196 \text{ m/s}$$

and $\nu^2/2g = 0.196^2/2(9.81) = 0.002$ m, which was negligible enough not to be considered in the first place.

Example 6.4

A 40 cm (15.74 in.) diameter galvanized iron pipe is available to connect the two reservoirs described in Example 6.3. How much discharge can be expected?

Solution

From Table 6.2 the data for galvanized iron pipes are

$$e = 1.5 \times 10^{-4} \text{ m}$$

$$v \cdot e = 1.5 \times 10^{-10} \text{ m}^3/\text{s}$$

$$g \cdot e^3/v^2 = 33$$

As before,

$$h_L/L = 0.025$$

and the required scale reading on the horizontal coordinate of Figure 6.3 is

$$0.025(33) = 8.25 \times 10^{-1}$$

The relative roughness is

$$e/D = (1.5/0.4)10^{-4} = 3.75 \times 10^{-4}$$

Entering the graph at the horizontal coordinate and turning at the given relative roughness, one finds that the vertical scale is

$$\frac{Q}{v \cdot e} = 2 \times 10^9$$

From here the discharge sought is given as

$$Q = 2 \times 10^9(1.5)10^{-10} = 0.3 \text{ m}^3/\text{s } (10.6 \text{ ft}^3/\text{sec})$$

which is less than the delivery of the larger and smoother steel pipe of Example 6.3.

Long before the Moody diagram made the results of modern scientific studies of pipe flow available for practicing engineers, there were many pipe flow formulas known. Most of these embodied experimental studies and their usefulness were limited to conditions similar to those of the experiments. While most of these old equations have fallen into disuse, one singular example of them still endures. This is the Hazen–Williams formula, shown in a nomographic form in Figure 6.4. The formula applies to water only. Its validity depends on the successful selection of the roughness coefficient C. Experience shows that for most smooth new pipes a value of $C = 140$ is reasonable. For rougher pipes, like concrete and vitrified clay, C is about 130. C values of 100 and less are generally applicable for old and badly corroded pipes.

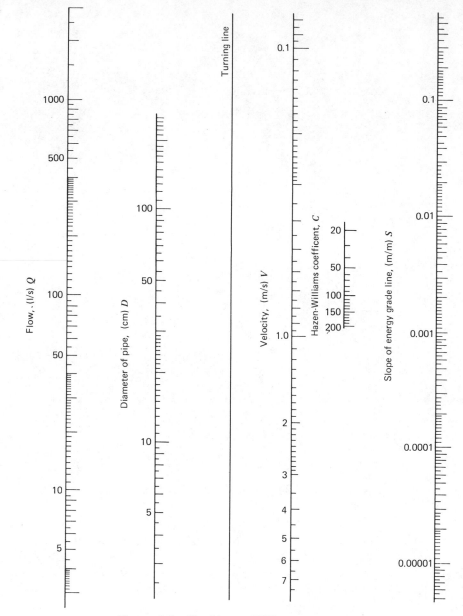

Figure 6.4 The Hazen–Williams nomograph.

6.3   Local Losses

In hydraulic practice pipes longer than 30 meters or so generate significantly more frictional losses along their length than the losses that are caused by inlet and exit conditions and fixtures along the pipe. Shorter pipes have relatively less frictional losses and in comparison local losses may become important.

Local losses in piping fixtures were found to be proportional to the kinetic energy, $v^2/2g$. The constant of proportionality is generally determined by laboratory experiments for each fixture type. On this basis the energy lost through a fixture may be computed by the formula

$$h_L = k \cdot v^2/2g \tag{6.6}$$

where k is the constant of proportionality. The k values of a number of common fixtures are listed in Table 6.3. In Equation 6.6 the velocity to be substituted is usually the velocity in the pipe before the fixture, unless otherwise specified in Table 6.3. In some cases it may be convenient to convert minor local losses into an "equivalent pipe length," which then could be added to the length of the pipe line considered. This may be accomplished by equating Equation 6.5 with Equation 6.6, resulting in

$$l_{equivalent} = \frac{k \cdot D}{f} \tag{6.7}$$

in which k is the proportionality constant from Table 6.3, D is the diameter of the pipe, and f is the friction factor.

Example 6.5

In the pipeline described in Example 6.2 there are the following fixtures contributing to additional energy losses:

strainer bucket with foot valve	(case 6 in Table 6.3)
close return bend	(case 14)
fully open gate valve	(case 16)
diffusor exit with 15 degree angle	(case 27)

Determine the additional pumping requirement due to these fixtures.

Table 6.3
Local Loss Coefficients

Use the equation $h_r = kv^2/2g$ unless otherwise indicated. Energy loss E_L equals h_r.

---

① Perpendicular square entrance:

$$k = 0.50 \quad \text{if edge is sharp.}$$

---

② Perpendicular rounded entrance:

$R/d =$	0.05	0.1	0.2	0.3	0.4
$k =$	0.25	0.17	0.08	0.05	0.04

---

③ Perpendicular reentrant entrance:

$$k = 0.8$$

---

④ Additional loss due to skewed entrance:

$$k = 0.505 + 0.303 \sin \alpha + 0.226 \sin^2 \alpha$$

Suction pipe in sump with conical mouthpiece:

⑤
$$E_L = D + \frac{5.6Q}{\sqrt{2g}D^{1.5}} - \frac{v^2}{2g}$$

Without mouthpiece:

$$E_L = 0.53D + \frac{4Q}{\sqrt{2g}D^{1.5}} - \frac{v^2}{2g}$$

Width of sump shown: $3.5D$

(After I. Vágás)

Strainer bucket:

⑥
$$k = 10 \quad \text{with foot valve}$$
$$k = 5.5 \quad \text{without foot valve}$$

(By Agroskin)

---

⑦ 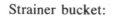 Standard Tee, entrance to minor line

$$k = 1.8$$

Table 6.3 (*Continued*)

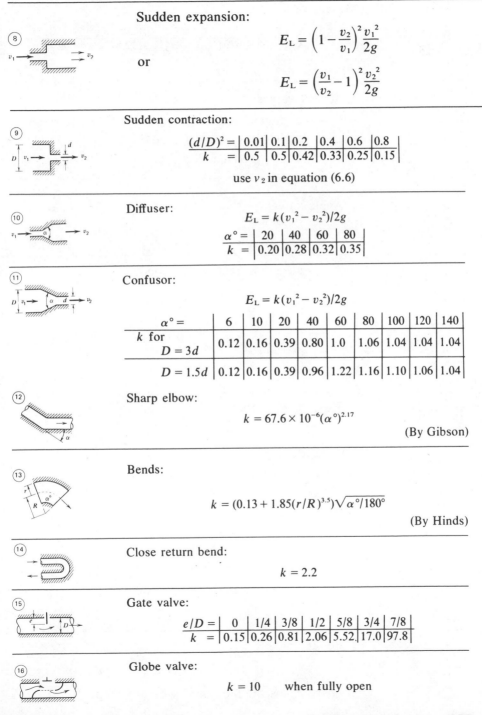

Sudden expansion:

$$E_L = \left(1 - \frac{v_2}{v_1}\right)^2 \frac{v_1^2}{2g}$$

or

$$E_L = \left(\frac{v_1}{v_2} - 1\right)^2 \frac{v_2^2}{2g}$$

Sudden contraction:

$(d/D)^2 =$	0.01	0.1	0.2	0.4	0.6	0.8
k =	0.5	0.5	0.42	0.33	0.25	0.15

use v_2 in equation (6.6)

Diffuser:

$$E_L = k(v_1^2 - v_2^2)/2g$$

$\alpha° =$	20	40	60	80
k =	0.20	0.28	0.32	0.35

Confusor:

$$E_L = k(v_1^2 - v_2^2)/2g$$

$\alpha° =$	6	10	20	40	60	80	100	120	140
k for $D = 3d$	0.12	0.16	0.39	0.80	1.0	1.06	1.04	1.04	1.04
$D = 1.5d$	0.12	0.16	0.39	0.96	1.22	1.16	1.10	1.06	1.04

Sharp elbow:

$$k = 67.6 \times 10^{-6}(\alpha°)^{2.17}$$

(By Gibson)

Bends:

$$k = (0.13 + 1.85(r/R)^{3.5})\sqrt{\alpha°/180°}$$

(By Hinds)

Close return bend:

$$k = 2.2$$

Gate valve:

$e/D =$	0	1/4	3/8	1/2	5/8	3/4	7/8
k =	0.15	0.26	0.81	2.06	5.52	17.0	97.8

Globe valve:

$$k = 10 \quad \text{when fully open}$$

Table 6.3 *(Continued)*

Rotary valve:

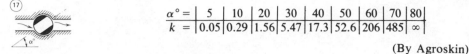

$\alpha° =$	5	10	20	30	40	50	60	70	80
$k =$	0.05	0.29	1.56	5.47	17.3	52.6	206	485	∞

(By Agroskin)

Check valves:

        Swing type $k = 2.5$ When fully open

        Ball type $k = 70.0$

        Lift type $k = 12.0$

Angle valve:

$$k = 5.0 \quad \text{if fully open}$$

Segment gate in rectangular conduit:

$$k = 0.8 \; 1.3\left[\left(\frac{1}{n}\right) - n\right]^2$$

where $n = \varphi/\varphi_0 =$ the rate of opening with respect to the ce[ntral] angle.

(By Abe[l]

Sluice gate in rectangular conduit:

$$k = 0.3 \; 1.9\left[\left(\frac{1}{n}\right) - n\right]^2$$

where $n = h/H.$

(By Bu[r]

Measuring nozzle:

        $E_L = 0.3 \, \Delta p$ for $d = 0.8D$

        $E_L = 0.95 \, \Delta p$ for $d = 0.2D$

where Δp is the measured pressure drop.

(By A.S.[M]

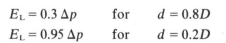

Venturi meter:

$$E_L = 0.1 \, \Delta p \quad \text{to} \quad 0.2 \, \Delta p$$

where Δp is the measured pressure drop.

Table 6.3 (*Continued*)

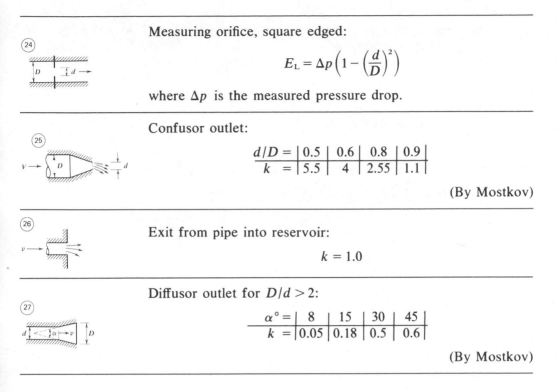

Measuring orifice, square edged:

$$E_L = \Delta p \left(1 - \left(\frac{d}{D}\right)^2\right)$$

where Δp is the measured pressure drop.

Confusor outlet:

$d/D =$	0.5	0.6	0.8	0.9
$k =$	5.5	4	2.55	1.1

(By Mostkov)

Exit from pipe into reservoir:

$$k = 1.0$$

Diffusor outlet for $D/d > 2$:

$\alpha° =$	8	15	30	45
$k =$	0.05	0.18	0.5	0.6

(By Mostkov)

Solution

Summing the local loss coefficients obtained from Table 6.3, one gets

$$k_{sum} = 10 + 2.2 + 10 + 0.18 = 22.38$$

The kinetic energy term, from Example 6.2, was

$$v^2/2g = 0.132 \text{ m}$$

Using Equation 6.6 the sum of the local losses is

$$h_{sum} = k_{sum}(v^2/2g) = 22.38(0.132) = 3.0 \text{ m}$$

Before local losses were determined, the total pumping requirement in Example 6.2 was 203.5 m. This must now be increased by three meters, that is, the pumping requirement shall be

$$H_{pumping} = 203.5 + 3.0 = 206.5 \text{ m} \ (677.3 \text{ ft})$$

Example 6.6

In Example 6.1 the 65-m long pipe starts from a standard Tee as its minor line (case 7), contains an open swing-type check valve (case 18), and exits straight into a reservoir (case 26 of Table 6.3). Express these local losses in equivalent length and recompute the total energy loss in the problem.

Solution

From Table 6.3 the sum of the local loss coefficients may be obtained as follows:

$$k_{sum} = 1.8 + 2.5 + 1.0 = 4.5$$

By Equation 6.7 the total equivalent length is

$$l_{equivalent} = k \cdot D/f = 4.5(0.25)/0.02 = 56.25 \text{ m}$$

Adding this to the actual pipe length of 65 m, one obtains a total effective length of 121.25 m. The energy loss, Equation 6.5, may now be recomputed as follows:

$$h_L = 0.02 \frac{121.25}{0.25} (0.032) = 0.31 \text{ m (1 ft)}$$

6.4 Slurry Flow

In dredging, in hydraulic transporting of soils for fills, and in moving fly ash or coal, pumping solid–water mixtures through pipes is often encountered. To enable the relatively heavy solid particles to be carried by fluid, there must be a significant degree of turbulence in the water. The vertically upward-acting turbulent velocity components will then cause shear forces on the solid particles that counteract their gravitational forces and allow them to be kept in suspension.

To compute the required pipe diameter and the energy losses that will occur when pumping slurries, the physical characteristics of the slurry must be known first. Specifically, the following three parameters must be determined:

The density of the solids carried: ρ_s (kg/m^3)

The statistical mean of the particle diameters, by weight: d (m)

The concentration or the percentage of solids in the liquid, by volume: c (percent/100)

If these parameters are given, the submerged specific gravity of the solids and the density of the slurry may be calculated by the formulas below:

Submerged specific gravity of solids:

$$S_{submerged} = \frac{\rho_s - \rho_{water}}{\rho_{water}} \qquad (6.8)$$

Slurry density:

$$\rho_{slurry} = (\rho_s) \cdot c + (1 - c)\,\rho_{water} \qquad (6.9)$$

As the density of water is unity for normal operating temperatures, the above two equations are rather simple. As a reminder, it may be mentioned that the density for sand may be taken as 2.6 g/cm³ or 2600 kg/m³.

Example 6.7

A slurry is composed of a mixture of sand and water. The mean diameter of the sand was determined by sieve analysis to be 1.5 mm (0.06 in.). The density of the sand is 2.64 g/cm³. The slurry concentration is 15 percent solids by volume. Determine the density of the slurry and the submerged density of the sand.

Solution

By Equation 6.8 the submerged specific gravity of the sand is

$$S_{submerged} = \frac{2.64 - 1.00}{1.00} = 1.64$$

The slurry density, with $c = 0.15$, and by Equation 6.9, is

$$\rho_{slurry} = 2.64(0.15) + 1.0(1 - 0.15) = 1.246 \text{ g/cm}^3$$

Theoretical and experimental studies of two-phase flow in pipes have led to Spells' equation that relates the slurry discharge to the pipe diameter required for a slurry of known physical characteristics. A modified and simplified version, using S.I. terminology, and applicable for water-based slurries, is

$$v = m \cdot (S_{submerged} \cdot d)^{0.816} (\rho_{slurry} \cdot D)^{0.633} \qquad (6.10)$$

In this equation D is in meters; v is in meters per second. The value m is 161 for the minimum velocity at which the slurry is not uniformly mixed across the pipe (denser on the bottom), but also at which the solids do not settle out; *m* is 475 for optimum velocity at which the slurry is homogeneously mixed throughout the pipe.

Example 6.8

A slurry described in Example 6.7 is to be pumped through a pipe with 0.25 m diameter (9.84 in.). By using the Spells' formula determine the minimum velocity of the flow at which the solids do not settle out and the optimum velocity at which the slurry is homogeneously mixed.

Solution

Substituting into Equation 6.10 one obtains

$$v = m(1.64 \ (0.0015))^{0.816} \ (1.246(0.25))^{0.633}$$

$$= m(0.00354)$$

For minimum velocity $m = 161$; therefore,

$$v_{minimum} = 0.57 \text{ m/s } (1.88 \text{ ft/sec})$$

For optimum velocity $m = 475$; therefore,

$$v_{optimum} = 1.68 \text{ m/s } (5.54 \text{ ft/sec})$$

Once the design velocity for the slurry is determined, a slurry pipeline is designed as though it would be carrying a common liquid. There are, however, two significant factors to be kept in mind: One is that the density of the fluid is heavier than that of water. It is computed by Equation 6.9; the result influences the magnitude of the Reynolds number by virtue of Equation 6.1. This, in turn, causes a change in *f*. Another factor in the design stems from the same difference in density, but it enters when the pump performance is considered. In determining the mass flow rate pumped, the slurry density should be included into Equation 7.1.

Example 6.9

The pipe considered in Example 6.8 is 350 m (1148 ft) long and is made of steel. What is the friction loss in the pipe when the slurry is pumped at minimum velocity?

Solution

The Darcy–Weisbach equation and the Moody diagram are utilized. The Reynolds number for the flow is

$$R = v \cdot D \cdot \rho/\mu = 0.57(0.25)\ 1.246/(10^{-6}) = 1.8 \times 10^5$$

The relative roughness is

$$e/D = 0.5 \times 10^{-4}/0.25 = 0.0002$$

Entering the Moody diagram one finds that

$$f = 0.017$$

The Darcy–Weisbach equation will then give

$$h_L = 0.017\ \frac{350}{0.25}\ \frac{0.57^2}{2(9.81)} = 0.39 \text{ m } (1.28 \text{ ft})$$

6.5 Branching Pipes

In practice, compound pipe flow problems are often encountered. By applying the continuity equation at each junction point and writing the Darcy–Weisbach equation for each of the interconnected pipes, such compound pipe problems can be solved. In order to demonstrate the method of solution, the most fundamental of these problems will be considered first: the problem of three interconnected reservoirs. As shown in Figure 6.5 three pipes leading from three reservoirs and connected together at a common junction point will result in a flow system between the reservoirs. The magnitude and the direction of the flow in each of the pipes will depend on relative elevation of the three reservoirs and on the energy losses through the pipes. By observing the probelm as shown in the drawing one will realize that:

a. the hydrostatic pressure at the junction point is a single value valid for each one of the pipes;
b. the sum of the discharge flowing in and out of the junction point must be zero.

From point (a) above it follows that the differences in piezometric heights between the junction point and at each reservoir define the head loss for each pipe. Equation 6.5 may be rearranged in the form of

$$\frac{8f}{g\pi^2 D^5} \cdot Q^2 = \frac{h_L}{L} = S \tag{6.11}$$

in which S is the slope of the energy line in the pipe considered. The first part of the left-hand side of Equation 6.11 depends on the pipe diameter and roughness only if one considers a fully turbulent flow in the pipe. Introducing a single variable to express this term we may write Equation 6.11 as

$$Q^2 / K^2 = S \tag{6.12}$$

where K is called the conveyance of the pipe and

$$K = \pi \sqrt{\frac{g \, D^5}{8 \, f}} \tag{6.13}$$

For fully turbulent flows in a steel pipe conveyance values are given for common pipe diameters in Table 6.4.

If the hydraulic energy level H at the junction is known, Equation 6.12 may be written for each one of the connecting pipes. For the three-reservoir problem the resulting equations would be

$$S_1 = \frac{H_1 - H}{L_1} = \left(\frac{Q_1}{K_1} \right)^2$$

$$S_2 = \frac{H_2 - H}{L_2} = \left(\frac{Q_2}{K_2} \right)^2 \tag{6.14}$$

$$S_3 = \frac{H_3 - H}{L_3} = \left(\frac{Q_3}{K_3} \right)^2$$

Depending on the relative magnitude of H, the values of S may be positive or negative depending on whether the flow at the branching point is toward or away from the junction.

Table 6.4
Conveyance Coefficients for Steel Pipe, Turbulent Flow Conditions

Nominal pipe diameter (in.)	Pipe diameter (cm)	Conveyance K (m³/s)
2	5.08	0.014
6	15.24	0.258
8	20.32	0.546
12	30.48	1.57
24	60.96	9.48

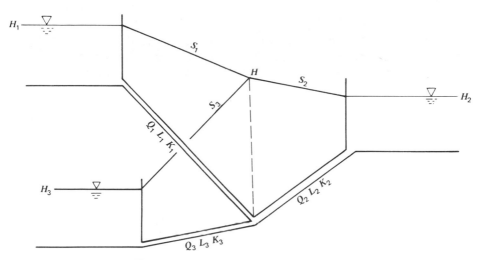

Figure 6.5 The three-reservoir problem.

The second consideration, (b), suggests that the equation of continuity must be satisfied at the junction. Selecting a sign convention such that discharges flowing toward the junction are positive, the equation of continuity results in

$$Q_1 + Q_2 + Q_3 = \Sigma Q = 0 \qquad (6.15)$$

Equations 6.14 and 6.15 when solved simultaneously provide a complete solution of the three-reservoir problem.

There could be a number of conditions for known or unknown parameters in such problems. The solution of any of these can be obtained by substituting the known terms into the preceding four formulas and solving them in a simultaneous manner. By the basic concepts of algebra, for the four equations we may have four unknowns. These problems are usually solved by trial and error.

The simplest application of the three-reservoir problem is that of the flow in two parallel pipes, shown in Figure 6.6a. The upstream branching point may be represented by two reservoirs of equal elevation. The head loss through both pipes will then be identical. The sum of the discharges flowing in the parallel pipes is the same as the discharge in the third pipe. Using the notation of Figure 6.6a, the formula solving a simple parallel pipe problem is

$$Q_0 = Q_1 + Q_2 = \sqrt{h_L}\left(\frac{K_1}{\sqrt{L_1}} + \frac{K_2}{\sqrt{L_2}}\right) \qquad (6.16)$$

in which h_L is the head loss between points A and B.

The principles used in the solution of branching pipes are also valid for the solution of pipe networks where there are scores of internal branching points. In municipal and industrial piping networks, many interconnected pipes are involved. The number of the resulting nonlinear algebraic equations to be solved is large. To solve the multitude of interdependent equations in practical applications, electronic computers are utilized. There are several computer programs available for the design and analysis of complex networks.

In Figure 6.6b a simple pipe network is shown. Each junction point A through K may be considered a three-reservoir problem. All of these problems are interdependent. A systematic trial and error procedure to solve such problems was introduced by Hardy Cross.

The first step in the application of the Cross method is to assume initial values for either all discharges in the individual pipes or all piezometric heads at the junction points. The second procedure will be explained below.

Denoting the assumed initial piezometric heads at junction points A, B, C, and so on, as H_A, H_B, H_C, . . . , the corresponding discharges can be calculated by Equations 6.14 because the head loss h_L for each pipe is given by the difference of heads at the junctions connected by the pipe in question. The conveyance of all pipes as well as their lengths are also known. After the discharges for all pipes are calculated, the continuity equation may be written for all junctions. For the initially assumed piezometric heads to be correct—a remote possibility—all continuity equations must be satisfied. Otherwise, an error in the continuity equations, ΔQ, must appear at each junction in the form of

$$\Sigma Q = \Delta Q \qquad (6.17)$$

This indicates that the assumed piezometric heads must be adjusted. The amount of adjustment, ΔH, is calculated in a systematic manner by the weighted formula

$$\Delta H = \frac{\Delta Q}{\Sigma |Q/h_L|} \qquad (6.18)$$

where the denominator is the sum of all incoming and outgoing discharges divided by their respective head losses in the connecting pipes. After all ΔQ values for the initially assumed piezometric heads are calculated, the required head corrections can be determined by Equation 6.18 at each junction point. For these new piezometric heads, new head loss and discharge values may then be computed. The process is repeated until the errors in discharges are reduced to an acceptable level.

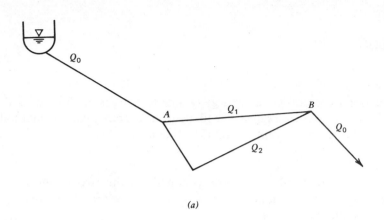

(a)

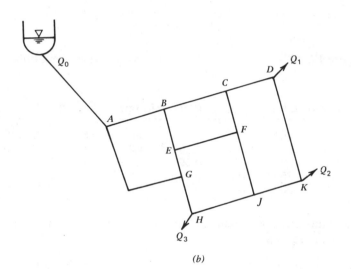

(b)

Figure 6.6 Piping systems: (a) parallel pipes; (b) pipe network.

Example 6.10

Determine the outflow discharge Q and discharges in all pipes in the pipe network shown in Figure E6.10 using the Hardy Cross method. The head loss between points 0 and C is 100 m.

Solution

First pressure head values are assumed for the two unknown internal junction points. Assume $H_A = 40$ m and $H_B = 60$ m. Accordingly, the heads at all points are

$$H_0 = 100 \text{ m}$$
$$H_A = 40 \text{ m}$$
$$H_B = 60 \text{ m}$$
$$H_C = 0 \text{ m}$$

From Figure E6.10, the following hydraulic parameters are tabulated:

Pipe	D (in.)	K (m³/s)	L (m)	$\sqrt{L}$	$K/\sqrt{L}$	h_L (m)	$\sqrt{h_L}$	Q (m³/s)	$|Q/h_L|$
0A	8	0.546	250	15.8	0.035	60	7.75	0.27	0.0045
0B	12	1.57	300	17.3	0.09	40	6.32	0.57	0.0142
AC	12	1.57	300	17.3	0.09	40	6.32	0.57	0.012
BC	8	0.546	250	15.8	0.035	60	7.75	0.27	0.0045
AB	12	1.57	200	14.1	0.11	20	4.47	0.49	0.0245

In this tabulation, the K values are from Table 6.4. The discharges are calculated using Equation 6.12.

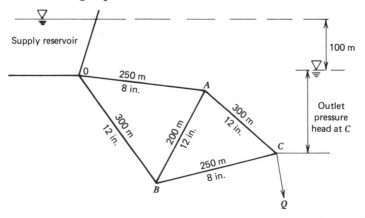

Figure E6.10 Pipe network showing lengths and nominal diameters.

The next step in the computations is the determination of excess flows at junctions A and B using Equation 6.17. These are:

$$\Delta Q_A = (0A) + (BA) + (AC) = 0.27 + 0.49 - 0.57 = +0.19 \text{ m}^3/\text{s}$$

$$\Delta Q_B = (0B) - (BA) - (BC) = 0.57 - 0.27 - 0.49 = -0.19 \text{ m}^3/\text{s}$$

Because the assumed pressures at points A and B are incorrect, the continuity equations are not satisfied at either point. Using Equation 6.18 the assumed heads are corrected as follows:

Point	$\Sigma|Q/h_L|$	ΔH	New H (m)
A	0.0432	+4.4	44.4
B	0.0432	−4.4	55.6

The computations shown in the table above may now be repeated to obtain a better value for the discharges (the first five columns will remain the same):

Pipe	h_L	$\sqrt{h_L}$	Q (m³/s)
$0A$	55.6	7.45	0.26
$0B$	44.4	6.6	0.59
AC	44.4	6.6	0.59
BC	55.6	7.45	0.26
AB	11.2	3.3	0.36

The computation is continued as before. Checking for excess discharges:

$$\Delta Q_A = 0.26 + 0.36 - 0.59 = 0.03 \text{ m}^3/\text{s}$$

$$\Delta Q_B = 0.59 - 0.36 - 0.26 = -0.03 \text{ m}^3/\text{s}$$

These relatively small errors may be neglected. The outflow discharge at point C will then be $Q_{AC} + Q_{BC} = 0.59 + 0.26 = 0.85 \text{ m}^3/\text{s}$.

Problems

6.1 Water at room temperature flows at an average velocity of 1.7 m/s through a pipe of 0.18 m diameter. Determine the Reynolds number of the flow. (*Ans.* 3×10^5)

6.2 A horizontal pipeline is connected to a reservoir. It consists of three pipe sections of 70 m each whose diameters are 6 cm, 14 cm, and 8 cm along the flow. The end of the pipe allows the water to exit into the air. Prepare a sketch showing the energy and hydraulic grade lines and label all losses that are encountered.

6.3 The head loss in a 200-m long 10-cm diameter steel pipe is 12 m. Determine the discharge using Asthana's diagram. (*Ans.* 0.01 m³/s)

6.4 The diameter of a 500-m long galvanized iron pipe is 0.2 m. The pipe carries 0.08 m³/s water. Determine the energy loss due to friction using the Darcy–Weisbach equation and Moody's diagram. (*Ans.* 15.8 m)

6.5 A steel pipe is 400 m long and connects two reservoirs whose elevations differ by 15 m. Determine the required pipe diameter to convey 0.24 m³/s discharge by gravity. (*Ans.* 25 cm)

6.6 A rotary valve is turned 40 degrees toward closure. How much is the loss of energy in it if the kinetic energy in the pipe is 1.2 m? (*Ans.* 20.76 m)

6.7 Express the local losses in Example 6.5 in the form of equivalent length. (*Ans.* 264 m)

6.8 A slurry contains quartz sand of 2 mm mean diameter with a concentration of 15 percent. If the slurry is to be moved in a 0.3 m diameter steel pipe, what should be the minimum discharge allowing homogeneous mixing along the pipe? (*Ans.* 0.21 m³/s)

6.9 Determine the friction loss if the pipe in Problem 6.8 is 300 m long.

6.10 Rework Example 6.10 with line 0A 24 in. in diameter and line *BC* 350 m long.

Chapter 7
Pumps

7.1 Types of Pumps

In the various fields of technology there is a huge variety of machinery built to transform hydraulic energy into mechanical energy and vice versa. Steam, gas, and hydraulic turbines generate mechanical energy from the kinetic, pressure, and potential energies of fluids. Fans transfer mechanical energy into moving air. Fluid couplings, like the automatic transmission of an automobile, utilize fluids as a transfer agent in altering mechanical energy. Pumps add energy to fluids. The common name for all of these devices is *turbomachinery*. There is a commonality in the theoretical and design concepts of turbomachinery and the field is an integral part of mechanical engineering. While people in hydraulics use certain types of pumps and hydraulic power generating turbines extensively, the mechanical design of these is clearly outside the field of hydraulics. Hence, in this chapter the internal design of pumps will be considered only to the extent of its proper application in hydraulic practice. In case of doubt the hydraulic designer should realize that pump manufacturers offer generous assistance to the users in order to assure the most efficient and proper application of their products. Turbines will not be discussed here at all. While they are interesting from the standpoint of general knowledge, few hydraulic engineers have a chance in their lifetime to work with them, and indeed, few will ever see one. Pumps, on the other hand, are often encountered.

Unless water is moved by gravity at an adequate discharge and pressure, it may be necessary to install pumps. There are many kinds of pumps used in the various technological fields, but the three main classes sort them out: centrifugal, rotary, and reciprocating. These classes refer to different ways pumps move the liquid. The different classes could be further subdivided into pumps of different types. For example, centrifugal pumps include the following types: propeller (axial flow), mixed flow, vertical turbine, regenerative turbine, diffuser, and volute.

This classification of centrifugal pumps is based on the way the rotating component—the impeller—imparts energy to the water. In turbine pumps or radial flow pumps the impeller is shaped to force water outward at right angles to its axis. In mixed flow pumps the impeller forces water in a radial as well as an axial direction. In propeller pumps the impeller forces water in the axial direction only. Within radial flow pumps we speak of a volute, diffuser, or circular type of casing, referring to the way water is collected and steered toward the exit pipe after it leaves the impeller.

Any one of the above types can be single stage or multistage pumps, where stage refers to the number of impellers in a pump. Another distinguishing characteristic is the position of the shaft, which can be vertical or horizontal. There are single suction and double suction pumps. Pumps could also be grouped by their construction materials: bronze, stainless steel, iron, and their various possible combinations. Material becomes important when corrosive liquids are handled.

There are many other ways in which pumps may move liquids. However, in the field of hydraulics, when we deal specifically with water, the most common pumps are centrifugal. Therefore our discussion will focus on these.

The selection of the type of pump for a particular service is based on the relative quantity of discharge and energy needed. To lift large quantities of water over a relatively small elevation—for example, in removing irrigation water from a canal and putting it onto a field—needs a different kind of pump than when a relatively small quantity of water is to be pumped to great heights—such as in furnishing water from a valley to a ski lodge at the top of the mountain. To make the proper selection for any application one needs to be familiar with the basic concepts of operation of the main types of centrifugal pumps.

The water enters the pump at the shaft that rotates the impeller. The impeller is a series of propellers, vanes, or blades that are arranged peripherally about the shaft and may or may not be held together by one or two circular plates. As the motor rotates the shift at high speed thus exerting a torque T, the water is whirled around as it enters the impeller. The angular velocity ω imparted to the water particles throws them outward onto the wall of the casing. The casing is built so that it leads the water toward the exit pipe either by vanes or by its gradually expanding spiral shape. By proper design of the casing the kinetic energy imparted to the water by the impeller gradually changes into pressure energy.

Impellers of large radius and narrow flow-passages transfer more kinetic energy per unit volume than smaller radius impellers of large water passage. Pumps designed so that the water exits from the impeller at a radial direction impart more centrifugal acceleration than those from which water exits axially or at an angle. Therefore, the relative shape of the impeller determines the general field of application of a centrifugal pump.

7.2 Head, Discharge, and Power Requirements

The typical internal shape of a centrifugal pump is shown in Figure 7.1. The pump shown is a single stage, double inlet pump with a volute-type (spiral-shaped) casing. Pumps of identical design are made in various sizes. Once the type of pump most suitable for a given application is selected, the size will be determined by the required discharge and by the dynamic head needed for its delivery. The discharge capacity of a pump is defined in volume flow rate Q. In current American practice this discharge is expressed in gallons per minute with 1 gpm = 6.31×10^{-5} m³/s. The approximate rated capacity of pumps may be roughly computed from the size of the pump exit; the exit diameter in meters multiplied by 0.06 gives the approximate discharge of the pump in m³/s. The rated pump capacity is obtained by the manufacturer by testing it with cold water and at optimum efficiency.

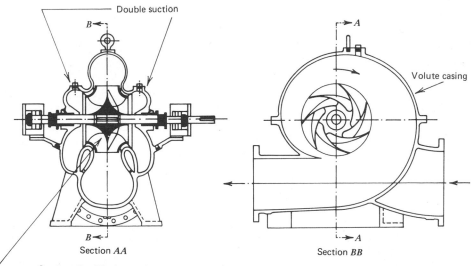

Figure 7.1 A typical cross-sectional drawing of a single stage, double suction, volute-type centrifugal pump.

The dynamic head H to be developed by a pump is computed according to the methods described in the previous chapter. The pressure or head to be developed is the measure of the height to which the water at the level of the reservoir or sump is to be lifted by pumping. In addition to this height the friction losses occurring in the pipe must be included. Losses of energy through the inlet devices such as strainer and foot valve, elbows, valves, and other components must also be taken into account. The pressure or kinetic energy required to be present at the end of the supply line is also a part of the *total dynamic head.* The speed of the impeller is expressed in its rpm (meaning rotations per minute). Although some pump motors are built such that the speed can be varied, this is a somewhat expensive and rare case. For most pumps standard electric motors are used. Standard speeds of alternating current synchronous induction motors at 60 cycles and 220 to 440 Volts are 3600, 1800, 1200, 900, 720, 600, and 514 rpm, depending on the number of poles. For a 50-cycle operation these speeds reduce to 3000, 1500, 1000 rpm and so on.

When selecting the motor for a pump and designing its wiring, fuses, and switching devices, it is important to know that pumps need more power for starting than for continuous operation. Since the full load speed of a regular electric motor is reduced by about 3 to 5 percent during startup, its requirement of power increases considerably.

Power required by a pump may be computed from the formula

$$P = \gamma \cdot Q \cdot H \qquad (7.1)$$

In this equation γ is the specific weight of the fluid pumped. For water this equals 9.81 kN/m³ (62.4 lb/ft³). The dimension of power P is meter · Newtons/second.

An electric motor's power requirement is usually stated in kilowatts. By the definition of the kilowatt, kN · m/s, it follows that if in Equation 7.1 P is in watts, then the right-hand side of the equation divided by 1000 gives the power in kilowatts.

In practice, the power needs of pumps are often expressed in horsepower. When defining the power requirement of a pump in horsepower, Equation 7.1 must be divided by 75. No pump will ever operate at 100 percent efficiency; therefore, the power of the driving motor must exceed the power needed by the water. The former is called *brake horsepower,* the latter is called *water horsepower.* They are related according to the formula

$$\text{brake hp} = \frac{\text{water hp}}{\text{efficiency}} \qquad (7.2)$$

Efficiency of a pump varies with Q and H. Also, different types of pumps have different typical efficiency values.

For each pump the manufacturer provides catalog information on its performance. This includes a graph showing the relationships between the discharge, the head, the brake horsepower, and the efficiency of the pump. One typical such performance graph is depicted in Figure 7.2.

There is a certain speed of revolution associated with each centrifugal pump at which the pump operates with the highest efficiency. For all other turning speeds the efficiency is less. Figure 7.3 shows an example of the changing of the head, discharge, and efficiency characteristics as a function of rotating speeds. The power can also be expressed as the product of the torque T acting on the shaft and the angular velocity ω of the rotating shaft, that is,

$$P = T \cdot \omega \tag{7.3}$$

The dimension of torque is (Newton $\cdot$ meter). The angular velocity of the pump's impeller is defined as

$$\omega = \frac{2\pi}{60} \cdot (\text{rpm}) \tag{7.4}$$

and its dimension is radians per second. If the shaft of the pump rotates with N rpm, the relationship between the torque T delivered and the power P required is

$$T = 9540P/N \tag{7.5}$$

where P is expressed in kilowatts.

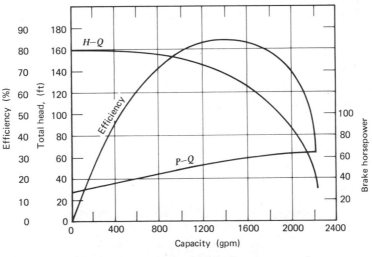

Figure 7.2 Typical pump performance graph.

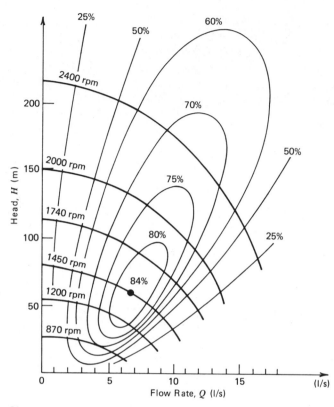

Figure 7.3 Characteristic curves for a centrifugal pump operating with different speeds. (Contour lines show constant efficiencies in percentage.)

Example 7.1

Determine the power required to deliver the flow in Example 6.2 if the efficiency of the pump is 80 percent.

Solution

The discharge pumped is 0.05 m³/s and the dynamic head neglecting local losses is 203.5 m. Using Equations 7.1 and 7.2 the energy required is

$$P_{\text{brake}} = \frac{\gamma Q \cdot H}{\text{efficiency}} = 9.81(0.05)203.5/0.8$$

$$= 124.6 \text{ kN} \cdot \text{m/s} = 124.6 \text{ kW}$$

Expressed in horsepower,

$$P = 124.6/0.746 = 167 \text{ hp}$$

Example 7.2

If the electric motor driving the pump of Example 7.1 is operated at 3600 rotations per minute (rpm), determine the torque acting on the drive shaft.

Solution

By Equation 7.4 the angular velocity of the shaft is

$$\omega = \frac{2\pi}{60} (\text{rpm}) = 0.105(3600) = 377 \text{ rad/s}$$

Using Equation 7.3 the torque T is then, multiplying by 1000 to get P in watts,

$$T = P/\omega = 124.6(1000)/377 = 330 \text{ N} \cdot \text{m}$$

Using Equation 7.5 directly, with P in kilowatts,

$$T = 9540(124.6)/3600 = 330 \text{ N} \cdot \text{m} \ (243 \text{ lb-ft})$$

Pump performance may be altered by changing either the impeller or the motor, or both. To change the pump performance characteristics certain basic laws valid for all centrifugal pumps are helpful. These are called *affinity laws* and are as follows:

Changing the impeller diameter D in the pump results in changes of Q, H, and P according to the relations

$$\frac{Q_1}{Q_2} = \frac{D_1}{D_2}$$

$$\frac{H_1}{H_2} = \left(\frac{D_1}{D_2}\right)^2 \tag{7.6}$$

$$\frac{P_1}{P_2} = \left(\frac{D_1}{D_2}\right)^3$$

Subscripts 1 and 2 refer to values of the parameters before and after the change, respectively. When only the motor speed is changed on the pump, the resultant changes follow the relationships below, as long as the pump continues to operate near its optimum efficiency:

$$\frac{Q_1}{Q_2} = \frac{N_1}{N_2}$$

$$\frac{H_1}{H_2} = \left(\frac{N_1}{N_2}\right)^2 \qquad (7.7)$$

$$\frac{P_1}{P_2} = \left(\frac{N_1}{N_2}\right)^3$$

where N_1 and N_2 refer to the rpm of the motors before and after the change. The similarity in Equations 7.6 and 7.7 shows that the change of impeller has the same influence on pump performance as the change of speed.

Example 7.3

If in Example 7.2 the rotational speed of the motor is reduced by 35 percent, what will be its effect on the pump's operation providing that the efficiency remains near its original 80 percent?

Solution

Since the original speed is

$$N_1 = 3600 \text{ rpm}$$

then by 35 percent reduction

$$N_2 = 3600(0.65) = 2340 \text{ rpm}$$

$$N_1/N_2 = 3600/2340 = 1.54$$

$$(N_1/N_2)^2 = (1.54)^2 = 2.37$$

$$(N_1/N_2)^3 = 3.65$$

From Example 7.1 the original performance values are

$$Q_1 = 0.05 \text{ m}^3/\text{s}$$

$$H_1 = 203.3 \text{ m}$$

$$P_1 = 124.6 \text{ kW}$$

Therefore, by Equation 7.7 the new performance characteristics are

$$Q_2 = 0.05/1.54 = 0.032 \text{ m}^3/\text{s}$$

$$H_2 = 203.5/2.37 = 86 \text{ m}$$

$$P_2 = 124.6/3.56 = 35 \text{ kW}$$

7.3 The Specific Speed

In Section 7.2 it was clarified that the performance of a pump depends on three parameters: the discharge, the dynamic head, and the speed of revolution of the impeller. In Section 7.1, furthermore, it was mentioned that pumps of identical shape are manufactured in different sizes and that the internal shape of pumps varies for different applications. To simplify the expression of the required pump characteristics the discharge, dynamic head, and speed of pumps may be consolidated into a single number called *specific speed*. The specific speed is evaluated at peak efficiency and is expressed by the formula

$$n_s = 284 \, \frac{NQ^{0.5}}{(g \cdot H)^{0.75}} \tag{7.8}$$

In this equation Q is to be substituted in m^3/s, g is 9.81 m/s^2, N is in rpm, and H is in meters. The multiplier, 284, accounts for the fact that in traditional American practice the specific speed was determined by the formula

$$n_s = (\text{rpm}) \, \frac{(\text{gpm})^{0.5}}{(\text{ft})^{0.75}} \tag{7.9}$$

which is not a dimensionless equation.

The specific speed of a pump is not really a speed in the physical sense, although it can be used in the sense that a pump reduced in size such that it would deliver one gpm to a height of one foot would run at its specific speed. In practice specific speed is only a number well suited to characterize the various types of centrifugal pumps. Generally, pumps with low specific speeds (500 to 2000 rpm) are made to deliver small discharges at high pressures. Pumps characterized by high specific speeds (5000 to 15,000 rpm) deliver large discharges at low pressures. The approximate relationships between specific speed, impeller shape, discharge, and efficiency are shown in Figure 7.4.

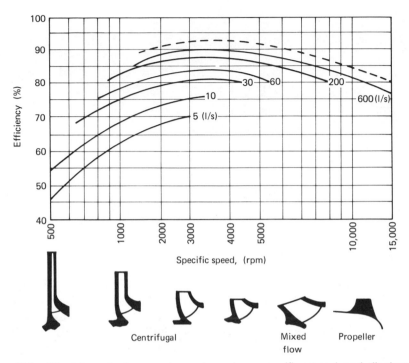

Figure 7.4 Chart to select pump types based on specific speed and discharge for optimum efficiency. (Lower drawings show impeller shape for the corresponding specific speed.)

The advantage of the use of Equation 7.8 stems from the fact that pumps built with identical proportions and shape but in different sizes have identical specific speeds. When the discharge, dynamic head, and required water horsepower is known, Figure 7.4 may be used to determine the type of pump needed.

Example 7.4

Select the type of pump best fitting the use described in Example 7.2. Determine the approximate value of the efficiency for these types of pumps.

Solution

First the specific speed is computed by Equation 7.8 with $N = 6300$ rpm, $Q = 0.05$ m³/s, and $H = 203.3$ m:

$$n_s = 284 \; \frac{3600(0.05)^{0.5}}{203.5^{0.75}} = 5741 \text{ rpm}$$

In Figure 7.4 the discharge of the pumps is shown in liters per second (l/s), and since there are 1000 liters in a cubic meter, we are seeking the 50 l/s region. It is in the upper part of the band between the 30 and 60 l/s lines. Following this band to the required value for specific speed (by extending the band toward the left), we find that for the discharge desired the efficiency will be below 85 percent. The figures on the bottom of the chart suggest that the type of pump will be of the mixed flow version. Except for the discharge in question the graph clearly indicates that for best efficiency a specific speed of about 2500 would be much better. Reviewing Equation 7.8 and keeping in mind that neither Q nor H could be changed, one concludes that reduction of the pump speed N is the answer. Either by rewiring or exchanging the electric motor, we can accomplish this. Of course, by virtue of Equation 7.3 the reduction of the speed will necessitate the increase of the torque in order that the power required to accomplish the task (124.6 kW) be maintained.

Once the type of pump is selected by evaluating the required specific speed, the pump manufacturers catalogs should be consulted for further information. Pumps of different discharge capacities are listed there under the same specific speeds. Particular attention is to be paid at that time to the performance curves like the one shown in Figure 7.1. The relative curvature of the pump characteristic lines and the relationship between head and discharge is quite different for different types of pumps. Axial flow pumps—those with high specific speeds—generally have steep characteristic curves. For radial flow centrifugal pumps—those having low specific speeds—the characteristic curves are usually flat. With flat characteristic curves, small changes of discharge result in small changes of head. With steep characteristic curves small changes of discharge may bring about large changes of head. This is a disadvantage of axial flow pumps when they are operated at less than peak efficiency. During startup, for example, their power needs may be much larger than under efficient, normal operating conditions.

All these considerations suggest that the selection of the best pump for the service required is not an easy task and as such it is best left to the customer service department of the manufacturer.

7.4 The Allowable Suction Head

An important point in the design of pumping installations is the elevation of the pump over the water level in the sump or reservoir from which the water is taken. The water in the suction line is in tension. The pressure is therefore lower than the atmospheric pressure. Adding to this already lowered pressure is the energy loss between reservoir and pump because of local losses and pipeline friction. The pressure is even more reduced since part of the energy at the pump is used in the form of kinetic energy because of the high velocities in the pump casing particularly around the impellers. This latter effect is related to the specific speed of the pump. Adding the elevation of the pump, the kinetic energy head, and the friction losses in the suction pipe of a pump, one obtains the *total suction head* H_s. If this total suction head corresponds to a pressure reduction in the pump that is equal to or less than the vapor pressure of the water (see Chapter 2), the water will change into vapor. This phenomenon is called *cavitation*. More than half of the troubles experienced with centrifugal pumps can be traced to the suction side and many of these problems involve cavitation. If the water vaporizes in the pump, small vapor bubbles form at the suction passages and at the impeller inlet. These bubbles collapse when they reach the region of high pressure. These collapses may occur with such violence that damage to the metal can result. Successive bubbles break up with considerable impact force, causing local high stresses on the metal surfaces that pit them along the grain boundaries of the casing and at the tips of the impeller. The presence of cavitation is easily recognized; the vibration and noise make the pump sound as though it were full of gravel. The result of cavitation is a significant drop in efficiency and a subsequent mechanical failure of the pump because of the cavitational erosion of the casing and the impeller and fatigue failure of the seals and shaft.

The allowable suction head in a given pump is the highest elevation above the downstream water level at which the pump will operate without a notable loss of efficiency due to cavitation. This height is expressed in terms of the total head H the pump is required to deliver by a factor of proportionality σ called *cavitation parameter*. The value of σ is determined by the manufacturer on the basis of the following simple test: The pump is set on a pedestal of adjustable elevation between two reservoirs. The total head H delivered by the pump is composed of the elevation difference between the reservoir levels, plus all energy losses, as shown in Figure 7.5. Measuring the brake horsepower and the discharge of the pump at various pump inlet elevations z, the efficiency of the pump may be calculated. As the pump's pedestal is raised beyond a certain elevation, the efficiency will begin to drop, indicating the onset of cavitation. This, in other words, means that p_o, the absolute pressure at the pump inlet, has been reduced to vapor pressure p_v. Using the notations of Figure 7.4 and writing the energy equation between the surface of the downstream reservoir and the pump's intake, the formula for the cavitation parameter may be derived as

$$\sigma = \frac{\dfrac{(p_0)_{abs}}{\gamma} - \dfrac{(p_v)_{abs}}{\gamma} - z - h_L}{H} \tag{7.10}$$

in which P_0 is the ambient atmospheric pressure, P_r is the absolute vapor pressure, h_L is the sum of the energy losses in the suction pipe, and H is the total dynamic head.

The numerator of Equation 7.10 is commonly called net positive suction head. H in the denominator is the total dynamic head. Values of the σ cavitation parameter range from 0.05 for a specific speed of 1000 to 1.0 for a specific speed of 8000. The value of σ is usually furnished by the manufacturer. For pumps of high specific speed, that is, low heads with large discharge capacity, the allowable net positive suction head may be less than

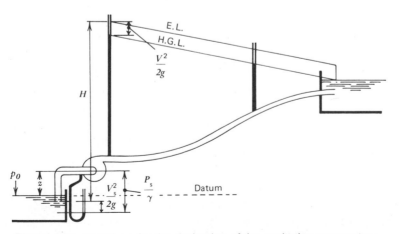

Figure 7.5 Notations for the derivation of the cavitation parameter.

zero, indicating that these pumps should be installed well below the reservoir water level in order to eliminate possible cavitation. In these instances the pump needs to be of the vertical shaft type so that the motor can be installed at an elevation above any possible flood level. Generally it was found that cavitation occurs as a result of too great a pump speed or too high a suction lift. Cavitation may also occur if the pump is not operating at or near the optimum point of its efficiency curve.

Example 7.5

For the problem described in Example 7.1 a pump is selected that has a cavitation parameter of 0.05. The losses in the suction pipe amount to 2.93 m (9.6 ft). For absolute ambient pressure 9.75 m (32 ft) of water may be assumed. The vapor pressure equals 0.24 m (0.79 ft). Determine the elevation of the pump with respect to the water surface of the sump.

Solution

Equation 7.10 may be rearranged to express the elevation difference z between the pump and the intake water level,

$$z = H_{\text{abs. atm.}} - (H_{\text{abs. vapor}} + h_L + \sigma H_{\text{dyn.}})$$

Substituting,

$$z = 9.75 - (0.24 + 2.93 + (203.5)\, 0.05)$$

$$z = 9.75 - 13.3 = -3.55 \text{ m } (-11.6 \text{ ft}) \text{ below sump level.}$$

7.5 Pumps in Parallel and in Series

In pumping stations the discharge and head requirements may fluctuate considerably in time. To operate at optimum efficiency the output of a single pump may fluctuate only within a rather narrow limit. Hence, for fluctuating needs it is advantageous to install several pumps in a pumping station. Pumps installed together may operate in series or in parallel. In either case the individual performance characteristics of the pumps must be carefully matched in order that the best overall efficiency be attained. It is advantageous to install pumps of identical size. In this case their matching is best from hydraulic standpoints. Also, from a practical standpoint the cost of stocking spare parts will be less and interchangeability of the various components will facilitate repair and maintenance. To eliminate downtime for repair, an additional pump of similar size may also be added to the system. This is put on line when any one of the other pumps needs repair or maintenance.

When two identical pumps are installed together in series, as shown in Figure 7.6, the total discharge of the two is the same as the discharge of a single pump, but the output pressure is doubled. When connected in parallel, the total delivered discharge of two identical pumps is twice that of a single

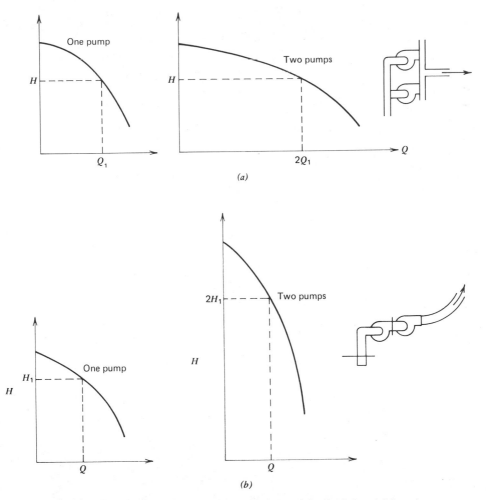

Figure 7.6 Two identical pumps operating in (a) parallel and (b) series.

pump, but the output pressure remains the same as the single pump. The above concepts are true only if the pumps operating together discharge into the atmosphere. When delivering water into a piping system that offers frictional resistance, two pumps operating in parallel will encounter greater resistance to flow. Hence they will have a different operating point than when they operate alone in the same piping system. Figure 7.7 illustrates this concept. The system's operating characteristic curve is determined here by selecting three or more discharges and computing the corresponding energy losses in the pipeline. In this figure the discharge versus head curve is plotted for the piping system. The operating points for the single or parallel operation of two pumps are also shown. The intersection of these lines are the operating points. As indicated in the graph, the joint discharge of two pumps in parallel is less than twice the discharge of a single pump. Similarly, two pumps operating in series will double neither head nor discharge in the whole system.

With these basic concepts, the joint operation of any arrangements of pumps can be analyzed. It requires the knowledge of the frictional resistances of the piping system and the availability of the characteristic curves of the pumps. The output of the whole system under various operating conditions can then be determined by graphical analysis.

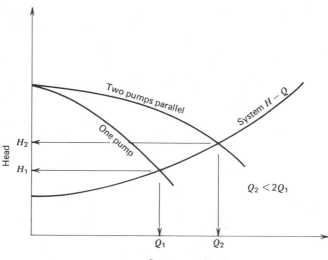

Figure 7.7 System performance involving two identical pumps operating in parallel and discharging water into a pipeline.

7.6   Starting and Troubleshooting

To begin the operation of a centrifugal pump it first should be primed, unless it is self-priming. *Priming* means to fill the pump with water so that the impeller can create suction. Foot valves serve the purpose of keeping the water in the pump between periodic operations, but often they leak. Priming requires that there be valves before and after the pump. These are to be closed before priming and starting the pump. After starting, the inlet valve is always opened first. After a short period, the outlet valve may be slowly opened. A closed outlet valve does not harm a centrifugal pump. With its outlet valve closed the pump pressure is increased by about 15 to 30 percent. The power load on the motor when the outlet is closed on the pump is reduced by about 50 to 60 percent.

Manufacture of pumps has reached such a level of sophistication that a pump is expected to give troublefree service for a long period of time. However, troubles may arise from improper design or installation or from poor maintenance. Some common problems and their probable causes are listed below:

If no water or not enough water is delivered:
　　Pump is not primed.
　　Speed is too low: check wiring.
　　Discharge head is too high.
　　Suction head is higher than allowed.
　　Impeller is plugged up.
　　Impeller rotates in the wrong direction.
　　Intake is clogged.
　　Air leak is in intake pipe.
　　Mechanical trouble (seals, impeller, etc.).
　　Foot valve is too small.
　　Suction pipe end is not submerged enough.
Not enough pressure:
　　Air is in the water through leak in suction pipe.
　　Impeller diameter is too small.
　　Speed is too low.
　　The valve setting is incorrect.
　　Impeller is damaged.
　　Packing in casing is defective.
Erratic action:
　　Leaks are in suction line.
　　The shaft is misaligned.
　　Air is in the water.

Pump uses too much power:
  Too high speed.
  Poorly selected pump.
  Water too cold.
  Mechanical defects.
Vibration and noise:
  Cavitation.
  Motor is out of balance.
  Bearings are worn.
  Propeller is out of balance (blade damaged).
  Suction pipe picks up air through vortex action in sump.
  Water hammer in the piping system.

Noise due to cavitation can be eliminated by allowing some air to enter the suction pipe; however, this will not eliminate cavitation.

Problems

7.1 Compute the water power needed to pump 0.15 m³/s of water to a height of 75 m. (*Ans.*   110 kW)

7.2 If the efficiency of the pump is 75 percent and the speed of the pump is 1800 rpm, what is the torque to which the drive shaft of the pump is subjected in Problem 7.1? (*Ans.*   781 N.m)

7.3 The impeller of a pump is changed from $D_1 = 26$ cm to $D_2 = 22$ cm. What will be the percentage reduction of the pump performance in terms of discharge, dynamic head, and power? (*Ans.*   1.18 Q, 1.39 H, 1.65 P)

7.4 For a discharge of 0.23 m³/s and dynamic head of 45 m, calculate the required rotational speed for a pump having a specific speed of 3500 rpm. (*Ans.*   2475 rpm)

7.5 Determine the specific speed if the discharge is 0.12 m³/s, the dynamic head of the pump is 68 m, and the speed is 3600 rpm. (*Ans.*   2698 rpm)

7.6 Using Figure 7.4 evaluate the type of pump needed for the conditions described in Problem 7.5. (*Ans.*   87% efficiency)

7.7 Assuming an absolute ambient pressure of 8.9 m of water and an absolute vapor pressure of 2.3 m of water, what is the permissible elevation of a pump with respect to the intake water level if the suction line's losses are 2.1 m and the total dynamic head is 40 m? The cavitation parameter is 0.1. (*Ans.*   0.5 m)

7.8   Prepare a sketch similar to Figure 7.7 for the system performance in the case of two pumps operating in series.

7.9   Why is it not recommended to operate two pumps in parallel or in series if the pumps are not identical?

7.10 A pump operates with a great deal of vibration and noise. List the probable problems.

Chapter 8
Open
Channels

8.1 Natural Conditions of Steady Flow

Because they both convey water, pipes and open channels have a lot in common. The primary differences between flow in pipes and flow in open channels are as follows:

Pipes	Open Channels
A. Flow is caused by pressure; therefore it will take place regardless of the alignment of the conduit.	Flow is caused by gravity; therefore the channel must have a downward slope.
B. The cross section of the conduit is constant along the pipe and defined by the diameter.	The channel cross section may vary along the path of the flow.
C. The pressure in the pipe, specifically at any point around the perimeter of the pipe, can be of any magnitude.	The perimeter of the cross section consists of two parts: the *free surface,* the water surface exposed to air, and the *wetted* perimeter where the water is in contact with the channel boundary. The pressure at the free surface is always zero (atmospheric).

Principles A, B, and C are sufficient to establish the fundamental equations of flow based on the concepts learned in Chapter 4. In this chapter the analysis of open channel flow problems will be limited to constant discharges. This requires the assumption that the discharge in the channel does not change in time; in other words, the flow is *steady*. This means that we exclude the problems of flood waves and tidal waves moving in the channel.

From the assumption of steady flow, the problem arising from principle B, listed above, becomes quite manageable. By the law of conservation of mass we may conclude that along the whole length of the channel the following equation must be valid:

$$Q = v_1 \cdot A_1 = v_2 \cdot A_2 = v_n \cdot A_n \tag{8.1}$$

In this equation Q is the constant discharge, v is the average velocity, A is the cross section of the flow, and the subscripts 1, 2, . . . , n refer to arbitrary positions along the path of the flow.

Example 8.1

The velocity at cross section 1 of an open channel is 1.5 m/s (4.9 ft/sec) and the cross-sectional area there is 3 m² (32 ft²). Some ways down the channel at cross section 2, the area perpendicular to the flow is 4 m² (43 ft²). What will be the corresponding reduction in the velocity and what is the discharge in the channel?

Solution

First the discharge may be computed by Equation 8.1, in which $v_1 = 1.5$ m/s and $A_1 = 3.0$ m². Hence the discharge is

$$Q = 1.5 \times 3 = 4.5 \text{ m}^3/\text{s}$$

Since $A_2 = 4.0$ m², the same equation would give us

$$v_2 = \frac{Q}{A_2} = \frac{4.5 \text{ m}^3/\text{s}}{4.0 \text{ m}^2} = 1.125 \text{ m/s (3.7 ft/sec)}$$

The velocity at cross section 2 could be expressed directly by writing the ratio from Equation 8.1 as

$$v_2 = v_1(A_1/A_2) = 1.5(3.0/4.0) = 1.5(0.75) = 1.125 \text{ m/s}$$

Principle C allows a great many types of flow conduits to be analyzed as open channels. Natural watercourses, like creeks, rivers, and mountain ravines, and man-made channels, like irrigation canals, drainage ditches, flood channels, and sewers are all open channels. The last one mentioned, the sewer line, is a special case. Sewers are commonly designed for gravity flow. When flowing full, the free surface in gravity flow is only a line along the top of the pipe. When the pressure at the top of the pipe exceeds the atmospheric pressure, open channel flow concepts do not apply anymore. Sewers operating under pressure, the so-called "force" sewers, are designed as pipes, using the principles described in Chapter 6.

The shape of the cross section of most man-made channels, including sewers, is most often uniform along the path of the flow. When the channel shape does not change along the path, we speak of *prismatic* channels. Even in the case of natural watercourses, one may determine the size and shape of an *average cross section* to simplify the handling of the computations. Often the assumption is made that the shape of the channel is rectangular. In streams the channel width, measured at the surface, greatly exceeds the depth. For such wide channels the introduction of the concept of *discharge per unit width,*

$$q = \frac{Q}{w} = \frac{\text{total discharge}}{\text{channel width}} \tag{8.2}$$

is often convenient. In channels of rectangular shape the change in the cross-sectional area A depends entirely on the change of depth y. In using the discharge per unit width, the area A can be replaced with depth y since $A = y \cdot 1$.

Example 8.2

A river is 30 m (98.4 ft) wide and its average depth is 1.8 m (5.9 ft). The discharge is 25 m³/s (882 ft³/sec). Determine the average velocity and the discharge per unit width.

Solution

The average velocity is

$$v = \frac{Q}{A} = \frac{Q}{w \cdot y} = \frac{25}{30 \cdot 1.8} = 0.46 \text{ m/s}$$

The discharge per unit width, by Equation 8.2, is

$$q = \frac{25 \text{ m}^3/\text{s}}{30 \text{ m}} = 0.83 \text{ m}^2/\text{s} \ (8.9 \text{ ft}^2/\text{sec})$$

Note that the average velocity could be calculated from q also by dividing with the depth y,

$$v = \frac{q}{y} = \frac{0.83}{1.8} = 0.46 \text{ m/s} \ (1.5 \text{ ft/sec})$$

In the previous calculations the term *average depth* was used. The average depth is defined as the cross-sectional area of the flow channel, measured in a direction perpendicular to the average velocity, divided by the surface width, that is,

$$y = A \,/\, w \qquad\qquad (8.3)$$

An alternate expression for this average depth, used by some authors, is "hydraulic depth." In the remainder of this chapter whenever the depth y is used in connection with channels that are not rectangular, we will mean average depth.

Another frequently used term in connection with open channels is the *hydraulic radius*. The hydraulic radius is defined as the ratio of the cross-sectional area and the wetted perimeter, or

$$R = A \,/\, P \qquad\qquad (8.4)$$

In principle C above, the wetted perimeter was explained as the length of the boundary of the cross-sectional area excluding the free surface, that is, the line along which the water is in contact with the channel material. It is clear that R is not a radius in its geometric sense. Rather, it relates two important parameters that influence the movement of the water in the channel. These are: the cross-sectional area, which is directly proportional to the quantity of the water carried in the channel, and the length of the solid surface where frictional forces resisting the movement are generated and that are inversely proportional to the quantity of discharge allowed to flow. The hydraulic radius, therefore, is a parameter indicating the relative transmissibility of a certain shape and cross-sectional area. For the same cross-sectional area A a wide and shallow channel would generate greater resisting forces than one where the depth is about the same as the width; hence the former would have less discharge carrying capacity. For each channel shape—rectangular, trapezoidal, parabolic, and so on—one would be able to find a certain geometric condition at which the hydraulic radius is optimal. The optimum hydraulic radius means that a given discharge can be carried at a given velocity with the smallest cross-sectional area. Figure 8.1 shows three design charts made for various side slopes that allow the determination of this optimum hydraulic radius for commonly used trapezoidal channels. The solid curves represent constant values of the hydraulic radius R. The straight lines emanating from the origin of the graphs indicate the location of the optimum hydraulic radius values and the region within which the hydraulic radius deviates from its optimal value by less than 3 percent. Other lines show constant bottom width versus water depth ratios. These are often used in design specifications. The term z in these figures indicates horizontal spread of the side slope of a channel for each meter of rise. This depends on the

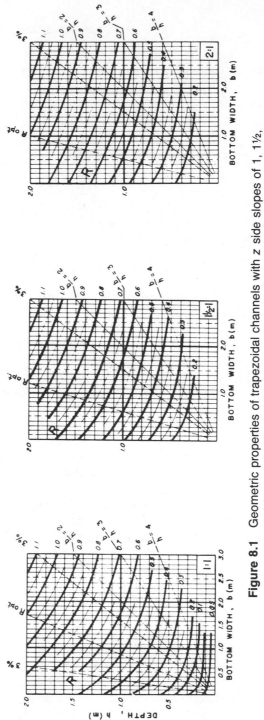

Figure 8.1 Geometric properties of trapezoidal channels with z side slopes of 1, 1½, and 2.

stability of the soil in which the channel is built. Recommended z values for different soils are listed in Table 8.1.

Table 8.1
Allowable Side Slopes for Trapezoidal Channels in Various Soils

Type of soil	z side slope
Loose clay or sandy loam	3
Loose silty sand	2
Firm clay	1.5
Earth with stone lining	1
Stiff earth with concrete lining	0.5–1
Muck and peat soils	0.25
Rock	0

In working with relatively shallow but wide channels, the hydraulic radius is often approximated by the average depth of the flow, that is, $R = y$. Another fact worthwhile to remember is that for pipes flowing full, the hydraulic radius equals one quarter of the diameter, that is, $R = D/4$.

Example 8.3

Determine the required depth and bottom width of a trapezoidal channel made in earth with stone lining if the required hydraulic radius is 0.8 m (2.6 ft). Show the range of depth and bottom width pairs for which the deviation from optimum conditions is equal to 3 percent, the values for optimum hydraulic radius, and the required size of the channel if the bottom width versus depth ratio is to be 2.

Solution

According to Table 8.1 the allowable side slope is 1. Hence the first graph (1:1) of Figure 8.1 is applicable. After locating the $R = 0.8$ line, we may read off the following depth and bottom width pairs. The 3 percent lines give $h = 1.9$ m with $b = 0.5$ m; and $h = 1.38$ m with $b = 2.2$ m. The $R_{\text{opt.}}$ is attained if $h = 1.5$ m and $b = 1.4$ m. The required $b/h = 2$ line is at $h = 1.3$ m (4.3 ft) and $b = 2.6$ m (8.5 ft).

Principle A introduced at the beginning of this section states that the driving energy in open channels is gravitational energy. If two points along the path of the channel are L distance away and the differences in elevation between those two points is H, then, when the water flows from one point to the other, it will gain an amount of potential energy equal to H. (The dimension of this energy is Newton-meter/kg of water flow.) For a unit length of the channel, the gravitational energy gained is expressible by

$$S = H / L \qquad (8.5)$$

The term S is a dimensionless quantity and it expresses the slope of the channel bottom for prismatic channels. Frictional forces developing along the channel surface, defined by the wetted perimeter and the length (P and L), tend to resist the movement of the water. These frictional forces depend on the velocity of the flow, the material of the channel surface, the hydraulic radius, and other factors. Jointly they cause the flowing water to dissipate some of its energy content as it flows from one point to the other along the length of the channel. The turbulent friction between water particles and the solid walls will convert some of the available hydraulic energy into heat energy, which, in turn, will dissipate in the surroundings. In any case, to get from one point to another the water will lose some energy, say, an amount equal to H^*.

In common natural cases the bottom slope of a watercourse varies according to the topographic and geologic conditions. Flat and relatively steep slopes follow each other along the path. Furthermore, the discharge of both natural and man-made channels is rarely permanent. Floods and low flow periods alternate. For different discharges the cross section of the flow channel will be different. The depth of flow, the hydraulic radius, and all other geometric properties will change along with them. With all these parameters varying from place to place and time to time, we have no reason to assume that while flowing from one point to another the energy gained by the water (H) will always equal the energy dissipated by the required movement (H^*).

If the energy required is less than the energy gained, that is, $H^* < H$, then the net energy remaining in the water will be greater. Conversely, if $H^* > H$, then the net energy remaining in the water will be less. In either case the results would be a change in the flow: with greater energy available the water would tend to accelerate; with less energy it would slow down. As long as the discharge is the same, an increase in velocity would decrease the cross-sectional area required. Conversely, a decrease in velocity would increase the area. A look at Equation 8.1 confirms these results. A certain discharge in a given channel may flow through in such a manner that there is no energy gained or lost; that is, $H = H^*$. Any other discharge in the same channel would result in either acceleration or deceleration; that is, $H \neq H^*$. Conversely, the given discharge under different conditions of slope, roughness, hydraulic radius, and so on would probably not flow through without gaining or losing some energy. In conclusion, one must realize that there are an infinite variety of ways water may flow through a channel, one of which, under some special conditions when all parameters are just right, is the case when there is neither acceleration nor deceleration. This unique condition, when the velocity of the flow is constant along the channel, is called *normal flow*.

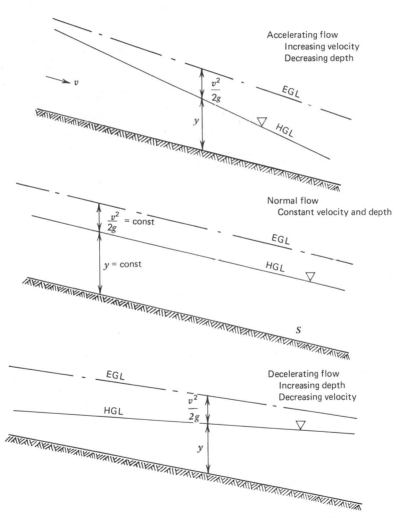

Figure 8.2 Channel bottom, energy grade line, and water surface in open channels; accelerating, decelerating, and normal flow. Steady discharge and prismatic channel are assumed.

8.2 Normal Flow

Since in normal flow the average velocity is constant, the kinetic energy, $v^2/2g$, does not change either. This implies, of course, that the flow in question takes place in a prismatic channel; that is, the cross-sectional area of the flow is constant. For this case the energy grade line is parallel with the hydraulic grade line, the free surface. The free surface, in turn, is parallel with the bottom slope of the channel, S, as shown in Figure 8.2.

The basic formula by which normal flow in open channels is calculated is the *Chézy–Manning equation*. The original Chézy formula takes the form of

$$v = C \sqrt{R \cdot S} \tag{8.6}$$

in which v is the average velocity of the flow, R is the hydraulic radius from Equation 8.4, S is the channel slope from Equation 8.5, and C is the so-called Chézy coefficient that accounts for the resistance of the channel. The last term was subject to much experimental and theoretical investigation. Several equations were proposed over the past century for the determination of C in terms of the channel roughness and channel geometry. The most commonly used one is Manning's equation,

$$C = \frac{1}{n} R^{1/6} \tag{8.7}$$

in which n is called the roughness coefficient and R, again, is the hydraulic radius. The roughness coefficient is often called the Manning's n; however, it was originally proposed by Kutter as a part of a slightly different formula for Chézy's C. The values of the roughness coefficient for various types of channels are listed in Table 8.2. While n appears to be a dimensionless constant for different channels, it is really neither dimensionless nor an independent constant. Its dimension is meter$^{1/2}$ per second. As a result, when Equation 8.7 is written in the conventional English system, the numerator 1.0 becomes 1.49. Studies in rocky channels have shown that n also depends on the velocity, the depth, and the slope of the channel. Figure 8.3 shows this dependence obtained by statistical analysis of field data.

The formula used for normal flow in open channels is obtained by combining Equations 8.6 and 8.7 and multiplying by the cross-sectional area A to form the Chézy–Manning equation,

$$Q = \frac{A}{n} R^{2/3} S^{1/2} \tag{8.8}$$

A nomograph to solve this formula is shown in Figure 8.4.

Table 8.2
Roughness Coefficients for Open Channels n

	n
Exceptionally smooth, straight surfaces: enameled or glazed coating; glass; lucite; brass	0.009
Very well planed and fitted lumber boards; smooth metal; pure cement plaster; smooth tar or paint coating	0.010
Planed lumber; smoothed mortar (⅓ sand) without projections, in straight alignment	0.011
Carefully fitted but unplaned boards, steel troweled concrete, in straight alignment	0.012
Reasonably straight, clean, smooth surfaces without projections; good boards; carefully built brick wall; wood troweled concrete; smooth, dressed ashlar	0.013
Good wood, metal, or concrete surfaces with some curvature, very small projections, slight moss or algae growth or gravel deposition. Shot concrete surfaced with troweled mortar	0.014
Rough brick; medium quality cut stone surface; wood with algae or moss growth; rough concrete; riveted steel	0.015
Very smooth and straight earth channels, free from growth; stone rubble set in cement; shot, untroweled concrete; deteriorated brick wall; exceptionally well excavated and surfaced channel cut in natural rock	0.017
Well-built earth channels covered with thick, uniform silt deposits; metal flumes with excessive curvature, large projections, accumulated debris	0.018
Smooth, well-packed earth; rough stone walls; channels excavated in solid, soft rock; little curving channels in solid loess, gravel or clay, with silt deposits, free from growth, in average condition; deteriorating uneven metal flume with curvatures and debris; very large canals in good condition	0.020
Small, man-made earth channels in well-kept condition; straight natural streams with rather clean, uniform bottom without pools and flow barriers, cavings and scours of the banks	0.025
Ditches; below average man-made channels with scattered cobbles in bed	0.028
Well-maintained large floodway; unkept artificial channels with scours, slides, considerable aquatic growth; natural stream with good alignment and fairly constant cross section	0.030
Permanent alluvial rivers with moderate changes in cross section, average stage; slightly curving intermittent streams in very good condition	0.033
Small, deteriorated artificial channels, half choked with aquatic growth, winding river with clean bed, but with pools and shallows	0.035
Irregularly curving permanent alluvial stream with smooth bed; straight natural channels with uneven bottom, sand bars, dunes, few rocks and underwater ditches; lower section of mountainous streams with well-developed channel with sediment deposits; intermittent streams in good condition; rather deteriorated artificial channels, with moss and reeds, rocks, scours, and slides	0.040
Artificial earth channels partially obstructed with debris, roots, and weeds; irregularly meandering rivers with partly grown-in or rocky bed; developed flood plains with high grass and bushes	0.067
Mountain ravines; fully ingrown small artificial channels; flat flood plains crossed by deep ditches (slow flow)	0.080

TABLE 8.2 (*Continued*)

Mountain creeks with waterfalls and steep ravines; very irregular flood plains; weedy and sluggish natural channels obstructed with trees	0.10
Very rough mountain creeks, swampy, heavily vegetated rivers with logs and driftwood on the bottom; flood plain forest with pools	0.133
Mudflows; very dense flood plain forests; watershed slopes	0.22

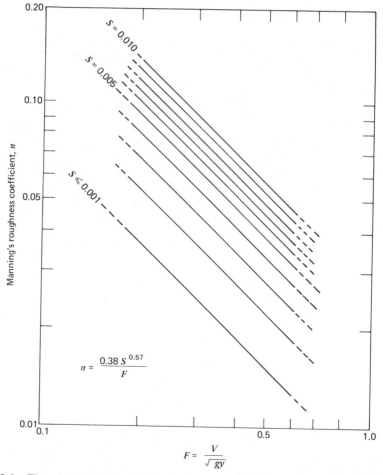

Figure 8.3 The dependence of Manning's *n* coefficient on the slope and on the Froude number in rough channels. (Courtesy of R. Garver.)

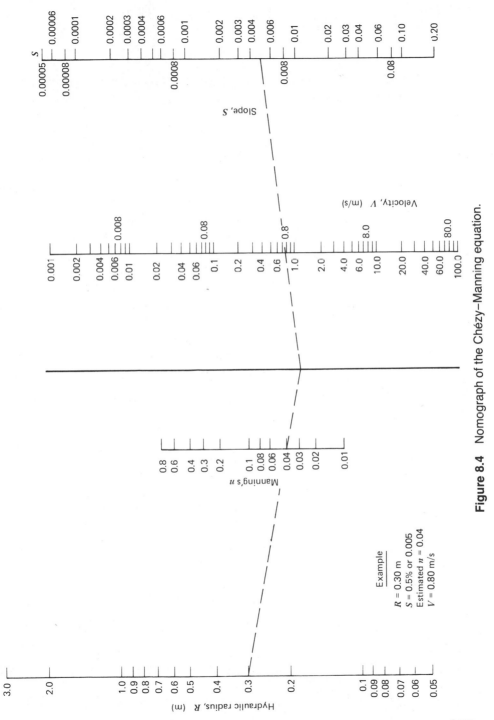

Figure 8.4 Nomograph of the Chézy–Manning equation.

153

Example 8.4

A 3-m wide rectangular concrete channel is laid on a $S = 0.005$ slope. How much is the discharge if the normal depth is 0.8 m?

Solution

The area is $A = 0.8 \times 3.0 = 2.4$ m². From Table 8.2 for concrete $n = 0.013$. The wetted perimeter is $P = 0.8 \times 2 + 3.0 = 4.6$ m. Hence the hydraulic radius (Equation 8.4) is $R = 2.4/4.6 = 0.52$ m. Substituting these values into Equation 8.8 we have

$$Q = \frac{2.4}{0.013} \, 0.52^{0.66} \, 0.005^{0.5} = 8.2 \text{ m}^3/\text{s} \ (290 \text{ ft}^3/\text{sec})$$

which is the answer sought.

Example 8.5

By using the Chézy–Manning nomograph shown in Figure 8.4, determine the increase of normal discharge in Example 8.4 in the case if the slope is increased to 0.015, assuming that the depth of the flow remains the same 0.8 m.

Solution

Because the depth will remain 0.8 m, the value of the hydraulic radius remains 0.52 m as well. This means that on the nomograph shown in Figure 8.4 the left solution line will intercept at $R = 0.52$ m and $n = 0.013$. Connecting the intercept on the centerline with 0.015 on the S line will cross the velocity line at $v = 6.05$ m/s. Since the cross-sectional area remained 2.4 m², the solution will be

$$Q_{\text{normal}} = A \cdot v = 2.4(6.05) = 14.58 \text{ m}^3/\text{s} \ (515 \text{ ft}^3/\text{sec})$$

Designers often face the problem of computing the normal depth for a given discharge. The difficulty here is that both A and R contain the unknown depth y in Equation 8.8. In order to solve such problems, Equation 8.8 must be rearranged by collecting all terms independent of the geometry on the left-hand side, that is,

$$\frac{n \cdot Q}{S^{1/2}} = A \cdot R^{2/3} = K \tag{8.9}$$

K in this equation is called the *conveyance* of the channel, which is a function of the geometry of the cross section. K for a given channel shape depends on the depth y alone. The most convenient way to solve Equation 8.9 for an unknown depth is to compute several values of K for pre-selected y's and plotting the results. Then the left-hand side of Equation 8.9 can be calculated, as it contains only known terms, and the plot may be entered with the resulting K to obtain the unknown depth y_n.

Example 8.6

A rectangular channel is 7.9 m (25.9 ft) wide and is cut in rock for which n is assumed to be 0.017. The discharge is 22 m³/s (777 ft³/sec). The channel is divided into three consecutive sections of varying slopes. In the first section $S_1 = 0.012$, in the second $S_2 = 0.05$, and in the last $S_3 = 0.006$. Determine the three normal depths, y_1, y_2, and y_3, using Equation 8.9.

Solution

Equation 8.9 will take the form of

$$\frac{n \cdot Q}{S^{1/2}} = y \cdot w \left(\frac{y \cdot w}{2y + w} \right)^{0.66} = K$$

To prepare the plot of K versus y we will select a series of depth values and compute

$$K = 7.9y \left(\frac{7.9y}{2y + 7.9} \right)^{0.66}$$

the results are

y (m)	0.1	0.5	1.0	2.0	5.0
$K = n \cdot Q/S^{1/2}$	0.158	2.29	6.8	19.1	66.9

The results are plotted in Figure E8.6.

The K values for the three different slopes are as follows:

$$K_1 = \frac{0.017(22)}{0.012^{1/2}} = 3.41$$

$$K_2 = 1.67$$

$$K_3 = 4.82$$

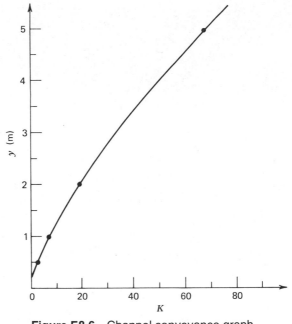

Figure E8.6 Channel conveyance graph.

Entering the plot of K versus y we obtain the three normal depth values sought as

$$y_1 = 0.6 \text{ m (2 ft)}$$

$$y_2 = 0.35 \text{ m (1.15 ft)}$$

$$y_3 = 0.77 \text{ m (2.5 ft)}$$

8.3 Critical Flow

There are two terms in the energy equation if one writes it with respect to the bottom of the channel: the kinetic energy, $v^2/2g$, and the pressure energy, y. At the end of Section 8.1 it was mentioned that there are an infinite variety of ways in which water may flow down a channel. However, the energy equation,

$$E = \frac{v^2}{2g} + y = \frac{Q^2}{2gA^2} + y \qquad (8.10)$$

controls the relationship between discharge, energy, velocity, cross-sectional area, and depth. Since the depth and the cross-sectional area are interdependent, there are three independent variables in Equation 8.10. These are the discharge, the total available energy, and the depth. In Figure 8.5 two graphs are shown: the D-shaped graph represents Equation 8.10 with the energy E held constant; the C-shaped graph represents the same equation when the discharge Q is held constant.

As the D-shaped graph indicates, a certain discharge Q' may flow through a channel section at two different depths, y_1 and y_2. These are defined by Equation 8.10 and referred to as *conjugate depths*.

When a dam's overflow structure discharges water from a lake, the amount of energy available for the flow is defined by the difference in elevation between the lake's surface and the weir's base line. In the case of the weir the discharge flowing will be the maximum possible under the available energy. This maximum is indicated by the peak of the D-shaped curve. It is called *critical discharge*. The singular depth value associated with the critical discharge is called *critical depth*. The velocity of the flow in the case described is referred to as *critical velocity*. Small surface waves, pressure waves (like sound waves in water), travel at critical velocity. The velocity of a pressure wave in water and other fluids is called *celerity*. The celerity c may be calculated in an open channel by the equation

$$c = \sqrt{g \cdot y} \qquad (8.11)$$

When water flows down a channel under steady conditions, the discharge is constant. Equation 8.10 takes the form of the C-shaped curve in Figure 8.5. Here the two variables are E and y. As indicated by the curve, a constant discharge may flow in an infinite variety of ways down a channel. For a given energy level, there are, again, two conjugate depths y_1 and y_2 at which the flow must take place. The only exception is the peak of the curve; it represents the minimum energy required to move the given discharge. For such minimum energy, the discharge and its associated depth are critical.

As it is known that the velocity of the critical discharge, $v_c = Q_c/A$, is equal to the celerity of the flow (Eq. 8.11), one may write the following relationship:

$$Q_c / A = \sqrt{g \cdot y_{ave}} \qquad (8.12)$$

in which y_{ave} is the hydraulic depth (Eq. 8.3). This equation may be rearranged such that all terms independent of the geometry are grouped on the left-hand side,

$$\frac{Q}{\sqrt{g}} = A \sqrt{y} = S \qquad (8.13)$$

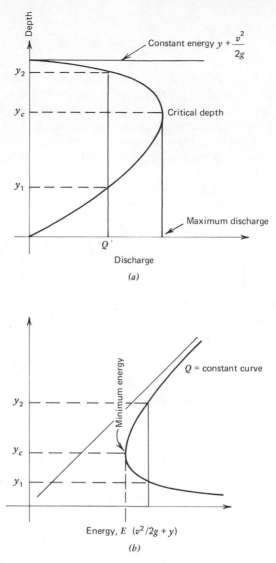

Figure 8.5 Graphical representations of the energy equation ($y + v^2/2g$ is the total available energy) for open channels. (a) Graph for constant energy, variable discharge, and depth. (b) Graph for constant discharge, variable energy, and depth.

158

where S is called the *shape factor*. Equation 8.13 is quite general and can be used to compute the critical depth for a given discharge. The method of computation is similar to the conveyance computation (Eq. 8.9), which was demonstrated in Example 8.6. First a series of arbitrary depth values are selected and their associated shape factors are computed. Plotting the shape factors against the depths results in a graph that allows the determination of the critical depth for any discharge flowing in the channel by the use of Equation 8.13.

Example 8.7

A 7.9-m (26-ft) wide rectangular channel carries 22.0 m³/s (777 ft³/sec) discharge. Determine the critical depth and the critical velocity.

Solution

By using Equation 8.13 the critical shape factor is computed to be

$$\frac{Q}{\sqrt{g}} = \frac{22}{\sqrt{9.81}} = 7.02$$

Next we select a series of arbitrary depths and compute the corresponding S values from the right-hand side of Equation 8.13,

$$S = 7.9y^{1.5}$$

The results are

y (m) =	0.1	0.5	1.0	2.0
S =	0.25	2.79	7.9	22.3

The tabulated values are plotted in Figure E8.7.

Entering the graph at the critical S value calculated above, one finds the critical depth to be

$$y_{\text{crit}} = 0.92 \text{ m (3 ft)}$$

The critical velocity is

$$v_{\text{crit}} = \sqrt{g \cdot y_{\text{crit}}} = 3 \text{ m/s (9.8 ft/sec)}$$

The critical flow condition separates two distinct ranges in which flow may take place in a channel. Maintaining the same discharge, if the velocity is less than the critical velocity, then one speaks of *subcritical flow*. The requirements of continuity, defined by Equation 8.1, state that the depth of subcrit-

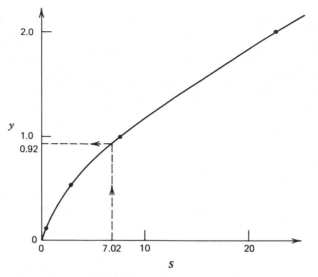

Figure E8.7 Shape factor versus depth.

ical flow must exceed the critical depth. If the velocity of the flow exceeds the critical velocity, the flow takes place in the *supercritical* range. In supercritical flows the flow depth is less than the critical depth.

The dimensionless parameter that indicates whether a flow is subcritical, critical, or supercritical is the *Froude number*. It is formed by dividing the average velocity of the flow with the celerity, defined by Equation 8.11, as follows:

$$\mathbf{F} = v / \sqrt{g \cdot y} \tag{8.14}$$

Using this definition, critical flows are represented by $\mathbf{F} = 1.0$, subcritical flows by $\mathbf{F}$ values below one, and supercritical flows by $\mathbf{F}$ values exceeding one.

Comparing the equation for normal discharge, Equation 8.8, with the equation for critical discharge, Equation 8.12, one may note that while the former is dependent on the slope of the channel, the latter is independent of it. One may equate the two discharges, substitute Q_{crit} into the place of Q_{norm}, and express the result in terms of the slope S. This slope is called *critical slope*. Obviously critical slope is strictly a hydraulic expression associated with one specific discharge flowing in a given channel. Such given discharge slopes that are flatter than critical are called *mild slopes*. On a mild slope the normal depth of the flow is greater than the critical depth and the Froude number associated with it will be less than one, signifying subcritical conditions. Conversely, when the slope at a given discharge is steeper than the critical slope, we speak of a *steep slope*. Normal flow on a steep slope is supercritical; the normal depth is less than the critical depth.

8.4 Erosion and Sedimentation

As water flows down an open channel, it usually carries a certain amount of sediment eroded from the channel banks or from its bottom. The sediment carrying capacity is directly proportional to the velocity of the flow. Sediment particles are suspended in the moving water by the vertical velocity components of turbulence. In reservoirs above dams the velocity and, in turn, the turbulence, are gradually reduced. This causes the sediments carried in the inflowing water to be deposited on the bottom of these pools. Similarly, as rivers enter the seas, their sediments are deposited forming big deltas like that of the Mississippi River. As the water passes through the dam, it regains its sediment carrying capacity. To satisfy it the river banks will erode below the dam until the water can pick up enough sediment corresponding to its carrying capacity. As the discharge changes before and after a flood peak passes, the sediment carrying capacity changes also. Even with steady flow, when the channel slope changes along the path of a river, the velocity of the flow will alter correspondingly. As a result, the wetted perimeter of a channel is rarely stable; it is constantly undergoing some change. Therefore the consideration of erosion and sediment deposition is an important part of hydraulic analysis.

Studies made by the U.S. Bureau of Reclamation indicate that erosion and scouring of earth channels will not occur unless the Froude number exceeds 0.35, that is,

$$v_{\text{eroding}} \geq 0.35 \sqrt{g \cdot y} \tag{8.15}$$

Of course, the inception of erosion will much depend on the soil in which the channel is built and the type of soil cover present.

To analyze the erosion potential of different soils, the so-called *tractive force method* may be applied. The tractive force of the moving water is computed by the equation

$$T = \gamma \cdot R \cdot S \tag{8.16}$$

where γ is the unit weight of the water (9.81 kN/m³), R is the hydraulic radius of the channel (m), and S is the channel slope. The allowable tractive force T for different soils is given in Table 8.3.

Table 8.3
Critical Tractive Force for Various Soils

Type of Soil	Critical tractive force (kN/m²)
Sand	1.7–4.0
Loose sandy clay	1.9–8.0
Firm clay	4.0–11.0
Muck and peat	2.0–12.0
Rock protection	150–250

Example 8.8

A rectangular channel cut in firm clay is 12 m (39.4 ft) wide and the depth of the water is 1.5 m (4.9 ft). The discharge is measured to be 23 m³/s (812 ft³/sec). The slope of the channel is 0.001. Assess the potential for erosion.

Solution

The cross-sectional area of the flow is $A = 1.5(12) = 18$ m². The velocity is therefore

$$v_{actual} = \frac{Q}{A} = \frac{23}{18} = 1.27 \text{ m/s}$$

Using Equation 8.15

$$v_{eroding} = 0.35 \sqrt{9.81(1.5)} = 1.34 \text{ m/s (4.4 ft/sec)}$$

Hence we may conclude that the velocity is perhaps low enough not to cause erosion. Next we check for tractive force, using Equation 8.16 as follows: First, the hydraulic radius must be computed from

$$R = \frac{A}{P} = \frac{18}{2(1.5) + 12} = 1.2 \text{ m}$$

Substituting into Equation 8.16,

$$T = 9810(1.2)0.001 = 11.7 \text{ kN/m}^2 \text{ (245.7 lb/ft}^2)$$

Table 8.3 lists the allowable tractive force for firm clay as ranging between 4 and 11 kN/m². Accordingly, the flow in our channel exceeds these limits; hence the channel will be eroding.

The two formulas introduced above set an upper limit on the flow velocity that may be allowed without the danger of scouring the channel. On the other hand there is a lower limit for the velocity of flow below which the suspended silt may settle out of the water. To prevent sediment deposition in open channels it is recommended that the design velocity exceed the amount corresponding to a Froude number of 0.12, that is,

$$v_{\min} = 0.12 \sqrt{g \cdot y} \qquad (8.17)$$

Example 8.9

Assuming the depth of flow is maintained as before, what would be the lowest limit of the discharge in Example 8.8 at which sedimentation would not occur?

Solution

Using Equation 8.17 with $A = 18 \text{ m}^2$,

$$v_{\min} = Q_{\min} / A$$

and

$$Q_{\min} = 18(0.12) \sqrt{9.81(1.5)} = 8.28 \text{ m}^3/\text{s} \ (292 \text{ ft}^3/\text{sec})$$

8.5 Backwater Concepts

Flow in an open channel is rarely uniform. Even if the discharge is steady, the stream decelerates or accelerates according to the geometric and hydraulic conditions present. One of the most common such conditions is the deceleration encountered in front of dams. Figure 8.6 shows the effect of a dam on water elevations upstream. The bottom of the stream channel is assumed to be straight, between points A and C. Without the dam the normal flow depth would be parallel with the bottom, represented by the line H to D. With a spillway height of C–E and no flow in the stream, the dam would create a pool of still water extending from E to B. If the discharge associated with the normal depth C–D would flow, it would pass over the dam at its critical depth, shown as E–F in the sketch. The velocity head corresponding to this critical depth is shown as F–J. If the shape of the weir at the top of the dam was known, both E–F and F–J could be computed. Such points along the channel where critical conditions are known to exist are called *control sections*. They are used as convenient starting points in determining water elevations upstream in their vicinity.

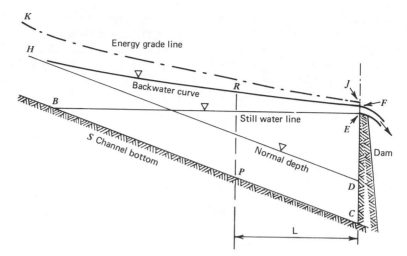

Figure 8.6 The effect of a dam on the water elevations upstream.

At the upstream end the influence of the dam gradually reduces to an insignificant amount, approaching the normal depth (*A–H*) in an asymptotic manner. The velocity head there is *H–K*. Directly in front of the dam, but not on the spillway, the velocity head is negligible because of the great depth of the flow, *C–F*. The line *H–F* represents the water surface under decelerating condition. It is called the *backwater curve*. Its computation in a precise manner is a very complex problem of hydraulics. There are at least a dozen different methods known to perform this computation, each with certain simplifying assumptions. Generally these methods assume that the flow is steady and the discharge is known, the slope of the channel is constant, and the shape of the cross-sectional area is defined as a rectangle, a parabola, a trapezoid, or other regular geometric shapes. Some of the simpler methods of computation are based on the additional assumption that the change in the velocity head (the difference in the curvature of lines *H–F* and *K–J*) is so small as to be negligible. One such simplified method was proposed over 100 years ago by Rühlmann. It assumes: the shape of the channel is a broad rectangle; the equation of Chézy (Eq. 8.8) is valid along the channel; the discharge *Q* is steady; the slope of the channel bottom *S* is constant; and the change of the velocity head along the channel is negligible. Because of its relative simplicity Rühlmann's method will be introduced in the remainder of this section. Because of the foregoing approximations this method may be used for preliminary work only. A more precise determination of the backwater elevations would require different methods and considerably more extensive calculations. When big dams are built on major rivers, the rise of the water surface requires relocation of highways and railroads, construction of miles of heightened flood control levees, relocation of utilities, rebuilding of bridges, raising sewer and drainage outlets often by requiring new pumping stations, and sometimes even the complete relocation of communities many miles upstream of the proposed dam. These changes often cost many millions of dollars. In light of these the expense of precise backwater computations is insignificant in practice. For small projects, or for preliminary estimates, the simple methods, like Rühlmann's, are quite sufficient.

Rühlmann's backwater equation is written in the form of

$$L = \frac{t}{S} \left[\psi\left(\frac{h}{t}\right) - \psi\left(\frac{z}{t}\right) \right] \qquad (8.18)$$

in which the notations, in terms of Figure 8.6, are as follows: L is the distance sought between two preassigned depths, namely z, which is P–R in Figure 8.6, and h, which is F–C in the deepest portion of the pool right at the dam. The term t is the normal depth A–H (or C–D), which may be computed by Equation 8.8. The functions ψ are shown in Table 8.4 for a wide range of h/t and z/t depth ratios.

The method of computation of the shape of the backwater curve proceeds in the following manner:

a.   First one determines the slope S, the normal depth t, and the depth h at the dam.

b.   Next a decision is to be made about the depth differentials in which the backwater curve needed to be known. Elevation differentials of 0.1, 0.25, or 0.5 m may be practical values. By gradually subtracting a standard depth-difference from the initial depth h, one arrives at a series of preassigned depths (P–R values) such as $z_1 = h - 0.25$ m, $z_2 = z_1 - 0.25$ m, $z_3 = z_2 - 0.25$ m, and so on until the final z_n value is close to the normal depth, t, signifying that the backwater effect of the dam is diminished to a negligible amount.

c.   By forming the ratios of h/t, z_1/t, z_2/t, and so on, one may obtain the functional values of $\psi(h/t)$, $\psi(z_1/t)$, and $\psi(z_2/t)$ from Table 8.4.

d.   Equation 8.18 may be entered with these values and a series of L values are computed that will correspond to the z depths.

e.   Finally the backwater curve is plotted using the calculated L_1z_1, L_2z_2, L_3z_3, . . . coordinates and connecting these points with a smooth line.

Table 8.4
Rühlmann's Backwater Function

$\frac{h}{t}$, $\frac{z}{t}$	$\psi\left(\frac{h}{t}\right)$, $\psi\left(\frac{z}{t}\right)$	Δ	$\frac{h}{t}$, $\frac{z}{t}$	$\psi\left(\frac{h}{t}\right)$, $\psi\left(\frac{z}{t}\right)$	Δ	$\frac{h}{t}$, $\frac{z}{t}$	$\psi\left(\frac{h}{t}\right)$, $\psi\left(\frac{z}{t}\right)$	Δ
0.01	0.0067	0.2377	0.36	1.4473	0.0165	0.92	2.1916	0.0233
0.02	0.2444	0.1419	0.37	1.4638	0.0163	0.94	2.2148	0.0232
0.03	0.3863	0.1026	0.38	1.4801	0.0161	0.96	2.2380	0.0231
0.04	0.4889	0.0812	0.39	1.4962	0.0157	0.98	2.2611	0.0228
0.05	0.5701	0.0675	0.40	1.5119	0.0156	1.00	2.2839	0.1132
0.06	0.6376	0.0582	0.41	1.5275	0.0155	1.10	2.3971	0.1113
0.07	0.6958	0.0524	0.42	1.5430	0.0153	1.20	2.5084	0.1095
0.08	0.7482	0.0451	0.43	1.5583	0.0151	1.30	2.6179	0.1085
0.09	0.7933	0.0420	0.44	1.5734	0.0150	1.40	2.7264	0.1073
0.10	0.8353	0.0386	0.45	1.5887	0.0148	1.50	2.8337	0.1064
0.11	0.8789	0.0359	0.46	1.6032	0.0147	1.60	2.9401	0.1057
0.12	0.9098	0.0336	0.47	1.6179	0.0145	1.70	3.0458	0.1050
0.13	0.9434	0.0317	0.48	1.6324	0.0144	1.80	3.1508	0.1045
0.14	0.9751	0.0300	0.49	1.6468	0.0143	1.90	3.2553	0.1041
0.15	1.0051	0.0284	0.50	1.6611	0.0282	2.00	3.3594	0.1037
0.16	1.0335	0.0273	0.52	1.6895	0.0277	2.10	3.4631	0.1033
0.17	1.0608	0.0261	0.54	1.7170	0.0274	2.20	3.5664	0.1030
0.18	1.0869	0.0250	0.56	1.7444	0.0270	2.30	3.6694	0.1026
0.19	1.1119	0.0242	0.58	1.7714	0.0266	2.40	3.7720	0.1025
0.20	1.1361	0.0234	0.60	1.7980	0.0263	2.50	3.8745	0.1023
0.21	1.1595	0.0226	0.62	1.8243	0.0260	2.60	3.9768	0.1021
0.22	1.1821	0.0219	0.64	1.8503	0.0257	2.70	4.0789	0.1019
0.23	1.2040	0.0214	0.66	1.8760	0.0254	2.80	4.1808	0.1018
0.24	1.2254	0.0207	0.68	1.9014	0.0252	2.90	4.2826	0.1017
0.25	1.2461	0.0203	0.70	1.9266	0.0251	3.00	4.3843	1.0115
0.26	1.2664	0.0197	0.72	1.9517	0.0248	4.00	5.3958	1.0062
0.27	1.2861	0.0193	0.74	1.9765	0.0245	5.00	6.4020	1.0036
0.28	1.3054	0.0189	0.76	2.0010	0.0244	6.00	7.4056	2.0041
0.29	1.3243	0.0185	0.78	2.0254	0.0241	8.00	9.4097	2.0023
0.30	1.3428	0.0182	0.80	2.0495	0.0240	10.00	11.412	5.002
0.31	1.3610	0.0179	0.82	2.0735	0.0240	15.00	16.414	5.001
0.32	1.3789	0.0175	0.84	2.0975	0.0238	20.00	21.415	10.000
0.33	1.3964	0.0172	0.86	2.1213	0.0236	30.00	31.415	20.001
0.34	1.4136	0.0170	0.88	2.1449	0.0234	50.00	51.416	50.004
0.35	1.4306	0.0167	0.90	2.1683	0.0233	100.00	101.420	

Example 8.10

A 10-m (32.8-ft) wide rectangular channel carries 12.73 m³/s (449 ft³/sec) discharge. The channel slope is 0.0017 and its roughness is $n = 0.022$. A dam at the downstream end of the channel raises the water level there to four meters above the normal depth. Determine the approximate back-water depths at three points along the channel.

Solution

Equation 8.8 may be used to compute the normal depth with the assumption that it equals the hydraulic radius. Hence

$$Q = \frac{w \cdot y_n}{n} y_n{}^{2/3} S^{1/2}$$

Substituting known values, one obtains

$$12.73 = \frac{10}{0.022} y_n{}^{1.66} 0.0017^{0.5}$$

Rearranging, one gets

$$y_n{}^{1.66} = t^{1.66} = \frac{12.73(0.022)}{(10) \, 0.0017^{0.5}} = 0.68$$

$$t = 0.68^{0.6} = 0.79 \text{ m}$$

Since the initial depth right above the dam site is $(t + 4)$ m, the value of h in Equation 8.18 is 4.79 m.

Next the decision may be made that one-meter depth differentials may be used. Hence the preassigned depth values will be

$$z_1 = h - 1 = 4.79 - 1.0 = 3.79 \text{ m}$$

$$z_2 = 2.79 \text{ m}$$

$$z_3 = 1.79 \text{ m}$$

The remainder of the computation, namely, the solution of Equation 8.18 for the given z values, may be done more conveniently in tabular form as shown below:

z (or h) (m)	z/t	ψ	$\Delta\psi$	$(t/S)\,\Delta\psi = \Delta L$	L (m)
4.79	6.06	7.41	—	—	0.0
3.79	4 79	6.18	1.2	557	557
2.79	3.53	4.92	1.3	604	1161
1.79	2.26	4.2	0.7	325	1486

$(t/S = 464.7 \text{ m})$

The right-hand column of this table shows the distances upstream where the preassigned depths shown at the left-hand side will occur.

Problems

8.1 Describe the essential differences between pipe flow and open channel flow.

8.2 What is a prismatic channel?

8.3 The bottom width of a trapezoidal open channel is two meters and the side slopes are 45 degrees. Compute the hydraulic radius for the case when the depth of the flow is 1.5 m. (*Ans.* 0.85 m)

8.4 What is the depth at which the hydraulic radius is optimum for the channel described in Problem 8.3? (*Ans.* 2.2 m)

8.5 The channel described in Problem 8.3 is laid on a slope of 0.005. The channel roughness is 0.05. Determine the normal discharge. (*Ans.* 6.6 m^3/s)

8.6 A rectangular channel is 5 m wide and carries a discharge of 3 m^3/s. The walls are made of concrete, and the slope is 0.008. Determine the normal depth. (*Ans.* 0.25 m)

8.7 Compute the critical depth for the case described in Problem 8.6. (*Ans.* 0.34 m)

8.8 Determine the Froude number in the case of Problem 8.5. (*Ans.* 0.33)

8.9 Compute the critical slope for Problem 8.6. (*Ans.* 0.0125)

8.10 A channel is to be laid on a slope of 0.005, built in firm clay, and expected to carry 3 m^3/s discharge. Design the channel in such a manner that neither scouring nor sedimentation would occur.

8.11 A channel is 15 m wide and of rectangular shape. The discharge is 5 m^3/s and the channel slope is 0.004. The roughness coefficient is 0.06. At the downstream end a dam causes the water level to rise to 4.5 m of pool height. Determine the approximate shape of the backwater curve.

8.12 Taking the case described in Problem 8.11 determine the upstream distance from the dam where the depth of the water would be twice the normal depth.

Chapter 9
Hydraulic Structures

9.1 Introduction

Hydraulic structures cause concentrated localized changes in open channel flow. While the flow in open channels may be considered analoguous to frictional losses in pipes, the analysis of the flow through hydraulic structures is comparable to that of local losses. A great variety of structures are built into open channels. Only a very small percentage of these follows a standardized design. The large majority of hydraulic structures are designed as unique installations. Specific needs of local hydrologic conditions, the differences arising from soil conditions, seepage, foundations, erosion, sedimentation and similar considerations, hydraulic operational requirements, and so on, prevent the introduction and application of standards.

Discharge in open channels is generally a random variable. It depends on the contribution rainfall over the watershed makes to the flow and on the many hydrologic characteristics that influence flood flow. The "design discharge" is an idealized concept subject to the caprices of nature. Any structure built into a watercourse will require a fundamental understanding of the principles of hydrology on the part of its designer. Questions like "what will happen during low flows," "what if the design discharge is exceeded," "what will be the effect of ice and freezing" should always be in the mind of the designer.

The literature related to the design of hydraulic structures is very large. Much of this material is scattered in reports, pamphlets, and internal specifications of various governmental and design organizations. There are only a few books in which design recommendations are collected and most of these specialize in certain types of structures. In spite of the great abundance of such information, the fluid mechanic fundamentals of the hydraulic design of structures do not extend beyond the material already learned in Chapters 4 and 8 of this book. The basic laws of conservation, local effects of friction, the concept of specific energy, and the concept of critical flow provide

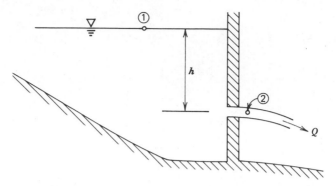

Figure 9.1   Notations for the orifice problem.

sufficient basis for establishing the design parameters required. Correction coefficients are introduced in the solutions to adjust to past experiences with similar types of structures. In the case of large, important, and expensive structures model studies are often performed in hydraulic laboratories. The experiments on these models are very economical substitutes for experience lacking with similar full-scale installations. Appendix 2 provides information about the design of these hydraulic models and about the interpretation of the result of laboratory studies.

Design formulas for hydraulic structures are generally made up of three major components. These consist of the fundamental fluid mechanics parameters, geometric parameters, and the parameters adjusting for turbulent and frictional effects. The latter in a simpler case is expressed in a *discharge coefficient*. The discharge coefficient is quite often a function of some of the geometric parameters.

As an example let us consider the case of flow through a small opening from a reservoir. In case a barrier is placed in a stream in which the flow takes place through a geometrically fixed opening located under the upstream water level, the flow is analyzed by the orifice formula. Consider a small opening called *orifice* as shown in Figure 9.1. The total available energy at the center of this opening equals the depth of the water h. Under the influence of this energy a jet of water will issue out of the orifice with a velocity v. Writing the energy equation for two points, one at the water surface of the container at point 1, another at the center of the jet at point 2, we obtain

$$\frac{v_1{}^2}{2g} + \frac{p_1}{\gamma} + Z_1 = \frac{v_2{}^2}{2g} + \frac{p_2}{\gamma} + Z_2$$

where the following assumptions may be made: Because of the large size of the container the approach velocity at point 1 is negligible when compared to the velocity of the jet at point 2, hence $v_1 = 0$. The pressure at point 1 as well as in the thin jet at point 2 equals the atmospheric pressure, hence, $p_1 = p_2$. Substituting $Z_1 - Z_2 = h$, the depth of the orifice below the surface, one gets

$$h = \frac{v_2^2}{2g}$$

or

$$v_{jet} = \sqrt{2gh} \qquad (9.1)$$

which is referred to as Torricelli's equation. The discharge through an orifice can be calculated if the area of the orifice A is sufficiently small with respect to the size of the container, in which case the variation of h from the bottom to the top of the opening is negligible. In this case the velocity of the flow throughout A can be considered constant and the discharge, at least theoretically, is

$$Q = A \cdot v = A \sqrt{2gh}$$

Experimental results showed that the actual discharge through an orifice is somewhat less than the discharge that would be obtained theoretically. There are two reasons for this discrepancy. First, the size of the jet is somewhat smaller than the size of the opening because of the effect of the curvature of the approaching stream lines in the container, as depicted in Figure 9.2a. The narrowest portion of the jet stream is called the "vena contracta." Another cause for the reduction of discharge is the viscous shear effect between the edge of the orifice and the water. This causes a loss of velocity around the edges, as shown in Figure 9.2b, with the corresponding decrease of discharge. The combined effect of these two factors gives rise to a *discharge coefficient, c,* whose values, based on many experiments with small, round, and square orifices discharging into air, range from 0.6 to 0.68. The experimental results indicate that the discharge coefficient is larger with smaller diameters and larger heads. Including the discharge coefficient c into the theoretical equation shown above, one obtains the discharge formula for flow through openings under pressure in the form of

$$Q = cA \sqrt{2gh} \qquad (9.2)$$

In the case when the jet issuing through the opening exits into the downstream water, the value of h in the above equation is the difference between the two water levels. For actual structures where the orifice cannot be considered infinitely thin but rather is a short tube, the value of the discharge coefficient increases. Figure 9.3 shows discharge coefficients based on experimental data on short tubes of various materials and configurations.

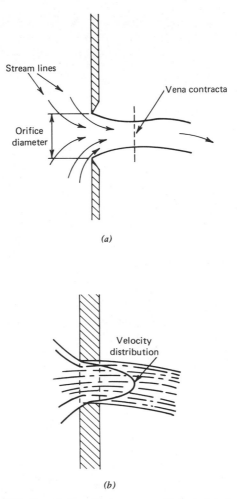

Figure 9.2 Effects contributing to the discharge coefficient: (a) the development of the "vena contracta"; (b) the velocity distribution due to viscous shear at the edge of the orifice.

In the remainder of this chapter three major kinds of hydraulic structures are considered in some detail. One, in which the motion of the water is caused by pressure; another in which the motion is entirely caused by gravity. These are the cases of flow under gates and over weirs, respectively. In the third kind of problem—flow through culverts—both pressure- and gravity-induced motion may be present. Most other hydraulic structures that are encountered in practice could be placed in any one of these classifications.

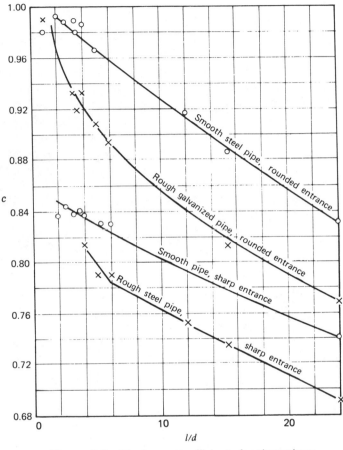

Figure 9.3 Discharge coefficients for short pipes.

9.2 The Hydraulic Jump

The velocity of flow through most hydraulic structures is often quite high. When the downstream water level is significantly below the upstream level, flow will take place without the retarding effect of the former. In such cases the flow will, at some point, reach critical conditions by virtue of the minimum energy theorem. Unencumbered by downstream influences, the flow over a weir will take place at critical velocity and will exceed critical conditions when it is shooting out from under a slightly raised gate. High velocities result in high frictional losses of energy. Supercrital flow below a gate will tend to slow rapidly. Decelerating supercritical flow results in a rise of water level. If the downstream water level represents subcritical flow conditions, the oncoming supercritical flow will suddenly turn subcritical when its depth reaches the conjugate value of the downstream water level. This sudden rise from supercritical to subcritical depth is called *hydraulic jump*. With the energy level and the discharge given, the corresponding conjugate depths, y_1 and y_2, may be determined by Equation 8.10. This equation may be simplified by considering a rectangular channel of unit width, where the cross-sectional area A may be substituted by the depth y and the discharge Q is replaced by $Q/(\text{width}) = q$. For practical use it is convenient to convert the formula by introducing the Froude number for the supercritical flow, **F**, as a variable in the form of Equation 8.14. The mathematical conversion results in a convenient dimensionless formula,

$$\frac{y_1}{y_2} = \frac{1}{2} \left(\sqrt{1 + 8F_1^2} - 1 \right) \tag{9.3}$$

Figure 9.4 shows these variables in connection with the hydraulic jump below the spillway of a dam. Depending on the magnitude of the Froude number at the beginning of the jump, we may note that the strength of the jump increases with F_1. For F_1 less than 1.7, experiments indicate that the jump appears as a series of undulating waves along the downstream surface. Between 1.7 and 2.5 a series of rollers appears on the surface, but the presence of the jump

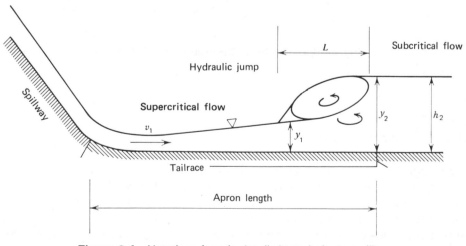

Figure 9.4 Notations for a hydraulic jump below a spillway.

does not significantly disturb the downstream water level and it is considered weak and unstable in its position. For Froude numbers of 2.5 and 4.5 the jump grows in strength, causing forceful random waves to travel downstream. For this range the elevation difference between the two sides of the jump is the greatest. For 4.5 to 9.0 the position of the jump stabilizes in the channel and is less influenced by the tailwater depth. For F_1 larger than 9.0 the jump again causes downstream waves, but their effect is less destructive because the energy dissipation in such jumps is high since a hydraulic jump is associated with high turbulence, vortices, and swirls. The shape and efficiency of a hydraulic jump for various Froude numbers is described by Figure 9.5. As a result there is considerable loss of total available energy in hydraulic jumps. The energy loss through a hydraulic jump may be computed by the formula

$$E_{\text{loss}} = \frac{[y_2 - y_1]^3}{4y_1 y_2} \tag{9.4}$$

The energy absorbing character of the hydraulic jump is very significant in the design of energy dissipators below hydraulic structures.

As the discharge and the two conjugate depths are inherently related by Equation 9.3, the relationship can be utilized to determine the discharge by

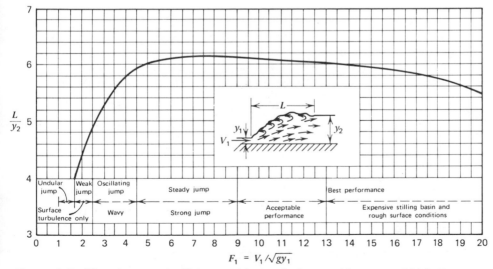

Figure 9.5 The shape and efficiency of hydraulic jumps. (Courtesy of U.S. Department of Reclamation.)

measuring the depths before and after the hydraulic jump in a rectangular channel and substituting into

$$q = \frac{Q}{B} = \frac{y_1 - y_2}{2} (gy_1y_2)^{1/2} \qquad (9.5)$$

where q is the discharge for a unit width of channel. There is no corresponding sudden change of water level when the flow changes from subcritical to supercritical. Hence there is no great energy loss associated with such changes. Hydraulic jump is only possible when the velocity is reduced from supercritical to subcritical.

As shown in Figure 9.4 the total length of the concrete apron below a spillway may only be partially covered by the jump. This happens particularly when the initial depth of the water is less than the lower conjugate depth of y_1, and the water will have to decelerate before entering the hydraulic jump. As building long aprons is a costly matter, designers strive to reduce their lengths as much as possible. Placing energy absorbing blocks and cross bars on the apron tends to increase the energy losses and hence aid deceleration. Another way of shortening the apron is to build a sill at the end of the tail race, forming a *stilling basin* in which the energy dissipates. The high tailwater elevation will force the jump to form against an increased downstream water level.

Example 9.1

On a 12-m (39.4-ft) wide rectangular chute water flows in supercritical condition at a rate of 150 m³/s (5295 ft³/sec). At the end of the chute on a horizontal concrete apron, the pool level of the downstream water is at 3 m (9.8 ft) above the apron. This downstream water will cause the formation of a hydraulic jump. Analyze the jump.

Solution

The discharge per unit width is

$$q = \frac{Q}{B} = \frac{150}{12} = 12.5 \text{ m}^3/\text{s}$$

The Froude number at the front of the jump, written in the form of Equation 8.14, is

$$F_1 = \frac{v}{\sqrt{gy_1}} = \frac{q/y_1}{\sqrt{gy_1}} = \frac{q}{\sqrt{g}\, y_1^{3/2}}$$

Using Equation 9.3 to express the conjugate depth y_1 for the known, $y_2 = 3$, we have

$$y_2 / y_1 = \frac{1}{2} \left(\sqrt{1 + 8F_1^2} - 1 \right)$$

$$= \frac{1}{2} \left(\sqrt{1 + 8q^2 / gy_1^3} - 1 \right)$$

From this, we get

$$y_1 y_2^2 + y_2^2 y_1 - 2q^2 / g = 0$$

and, substituting all known values,

$$3y_1^2 + 9y_1 - 2(12.5)^2 / 9.81 = 0$$

or

$$y_1^2 + 3y_1 - 10.62 = 0$$

Solving this quadratic equation, we find that

$$y_1 = 2.09 \text{ m}$$

Substituting into the expression for the Froude number, we get

$$F = \frac{q}{\sqrt{g}\, y_1^{3/2}} = \frac{12.5}{\sqrt{9.81}\, (2.09)^{3/2}} = 1.32$$

which is less than 1.7. The hydraulic jump will take the form of a series of undulating waves along the downstream water surface. The location of the jump is rather ill defined and shore protection is required due to the wave action.

The energy loss in the jump may be computed from Equation 9.4 as

$$E_{loss} = \frac{(y_2 - y_1)^3}{4y_1 y_2} = \frac{(3 - 2.09)^3}{4(3)2.09} = 0.03 \text{ m}$$

9.3 Flow under Gates

Flow under a *vertical gate* can be defined as a square orifice problem as long as the opening height, *a,* under the gate is small when compared to the upstream energy level H_0, and the downstream water level H_2 does not influence the flow. By Equation 9.2 we may write

$$Q = bac \sqrt{2g(H_0 - H_1)} \tag{9.6}$$

in which b is the width of the gate and the other terms are as shown in Figure 9.6. Direct use of this equation is difficult in practice because of the uncertainty in the determination of H_1, the depth of water at the vena contracta. Since H_1 depends on the opening height a, we may write

$$H_1 = \psi a$$

Experimental values for ψ were found to depend on H_0/a. For a values not exceeding $\frac{1}{2}H_0$, the value of ψ is about 0.625. The discharge coefficient c ranges from 0.59 for $a = \frac{1}{2}H_0$ to 0.607 for $a = 0.2H_0$. Equation 9.6 is valid only if the downstream water level is not influencing the flow. This is true when a hydraulic jump exists below the gate. This requires that the conjugate depth of H_2 downstream water level be equal to or larger than H_1.

Figure 9.7 gives the relationship between H_0, H_2, and a at the condition when the free outflow becomes retarded by the downstream water level.

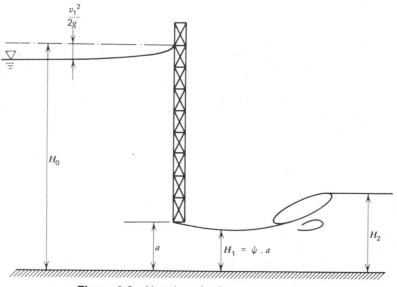

Figure 9.6 Notations for flow under gates.

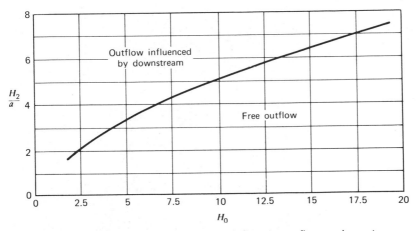

Figure 9.7 The range of downstream influence on flow under gates.

When $H_0 - H_1$ is equal or less than H_2, one speaks of *absolute downstream control*. In this case H_1 is completely submerged and the discharge computation is based on $h = H_0 - H_2$. In this case the discharge formula is

$$Q = bac \sqrt{2gh} \tag{9.7}$$

A nomograph combining all three possible cases, namely, free outflow, partial, and absolute downstream control, is shown in Figure 9.8. This graph allows the general solution of problems related to flow under gates.

Example 9.2

The upstream water level is 2.5 m (8.2 ft) and the downstream water level is 1.2 m on a 2-m wide vertical gate. How much will be the discharge if the gate is raised by 0.4 m?

Solution

Figure 9.8 may be used with

$$B = 2 \text{ m}, \quad H_0 = 2.5 \text{ m}, \quad H_2 = 1.2 \text{ m}, \quad a = 0.4 \text{ m}$$

The dimensionless variables are $a/H_0 = 0.4/2.5 = 0.16$ and $H_2/H_0 = 1.2/2.5 = 0.48$. Entering the graph we find that the flow will take place without downstream influence. This may be further checked by

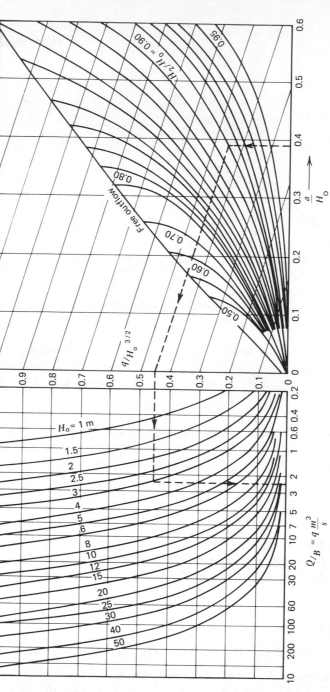

Figure 9.8 General solution for flow under gates. (After Chertousov. Courtesy of Ö. Starosolszky.)

entering Figure 9.7 with $H_0/a = 2.5/0.4 = 6.25$ and $H_2/a = 1.2/0.4 = 3.0$, again indicating free outflow. Completing the solution with Figure 9.8 we find that $Q/B = 1.8$, and since $B = 2$ m the discharge flowing will be $Q = 2(1.8) = 3.6$ m³/s (950 gal/sec).

9.4 Flow over Weirs

Flow taking place over a hydraulic structure under free surface conditions is analyzed with the *weir formula*. Generally speaking all barriers on the bottom of the channel that cause the flow to accelerate in order to pass through can be considered as weirs. More especially, weirs are constructed with openings of simple geometrical shapes. Rectangular, triangular, or trapezoidal shapes are the most common. In each case the bottom edge of the opening over which the water flows is called the *crest,* and its height over the bottom of the reservoir or channel is known as the crest height. The French term *nappe* (sheet) is often applied to the overfalling stream of water. Weirs where the downstream water level is below the crest allow the water to pass in a free fall. Under this condition weirs are good flow measuring devices, particularly if their wall at the crest and sides is thin. Weir shapes other than rectangular are almost exclusively used in flow measurement. The use of *sharp-crested weirs* in flow measurement was discussed in Chapter 5. For practical purposes as long as their crest thickness is more than 6/10 of the nappe thickness, weirs should be considered broad-crested. Flow over *broad-crested weirs* is significantly influenced by viscous drag, which causes a boundary layer to form in the velocity profile of the overflow. This effect is enumerated in the form of a discharge coefficient depending on the shape of the crest, and on the upstream energy level. If the width of the upstream channel is greater than the width of the weir opening, we speak of *contracted weirs*. The side contraction in weirs results in an additional contribution to the discharge coefficient by reducing the flow further. In weirs whose width equals the width of the upstream channel surface the side contraction is suppressed. These weirs are called *suppressed weirs*. For a rectangular weir with freely falling water the velocity at any point over the crest can be determined by the energy equation. The variables of the problem are shown in Figure 9.9. The total available energy on the upstream side is

$$E_{upstream} = h_1 + v_1^2 / 2g \qquad (9.8)$$

where h_1 is the depth of water on the upstream side over the crest back at a point where the water level is unaffected by the surface curve, and v_1 is the approach velocity in the channel. When the P height of the crest is significant,

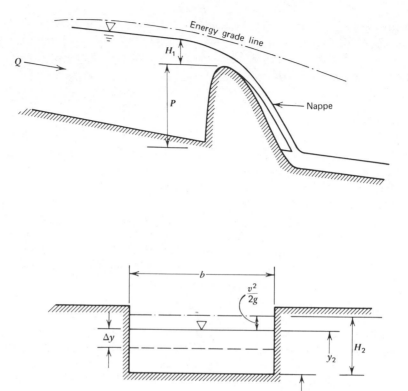

Figure 9.9 Notations for the weir formula.

as in most hydraulic structures, the approach velocity is relatively small; hence the kinetic energy on the upstream side may be neglected. For free fall the discharge Q will occur under critical flow condition. The depth of the nappe will be the critical depth, that is,

$$y_2 - y_1 = \tfrac{2}{3} E_{\text{upstream}}$$

or, when neglecting the approach velocity and considering $P = y_1$ as our datum plane,

$$y_2 - y_1 = \tfrac{2}{3} h_1 \tag{9.9}$$

The velocity at any point between y_2 and y_1 will be determined by the energy available at that point. For example, in a thin horizontal strip of just below y_2, the velocity equals

$$v = \sqrt{2g(h_1 - y_2)}$$

For the area of the thin strip of width Δy, shown in Figure 9.9, the elemental discharge is

$$\Delta q = b(\Delta y)v \simeq b(\Delta y)\sqrt{2g(h_1 - y_2)}$$

Summing the whole area of the nappe composed of many thin strips between y_2 and y_1 can be performed by a simple step of integral calculus. The result is the common form of the weir formula,

$$Q = cb\sqrt{2g}\,h_1^{3/2} \tag{9.10}$$

in which Q is the total discharge over the rectangular weir of b width under a head of h_1. The parameter c in this formula is the discharge coefficient of the weir. Equation 9.10 may be solved in metric terms by the aid of the graph shown in Figure 9.10. The formula plotted combines the value of c with the metric value of $\sqrt{2g} = 4.43$ in the form of

$$M = c\sqrt{2g} = Q / bH^{3/2} \tag{9.11}$$

where M is called the *metric weir coefficient*. The actual value of c in practice is generally determined by field tests. It ranges between 0.35 and 0.6 for most broad-crested weirs, depending on the depth of flow and the breadth of the weir. The broader the crest and the shallower the flow, the smaller is the discharge coefficient.

When the downstream water level h_2 exceeds the crest height, it may influence the discharge over the weir. In this case the water is prevented from passing by free fall. There are two such conditions. As long as the downstream water level is below the midpoint of the nappe over the crest, a hydraulic jump will form over a weir (see Fig. 9.11a). The nappe in this case will consist of a supercritical portion preceding the hydraulic jump. With further increase of the downstream water level the jump will first become submerged. With additional increase of h_2, the downstream water level will cause a rise in the upstream water level; the water will be backed up. For this case, shown in Figure 9.11b, the discharge could still be computed by Equation 9.10 with the stipulation that h_1 in the equation be replaced by the difference in the measured upstream and downstream water levels. In all cases when the free fall over the weir is prevented by the height of the downstream water level, the discharge coefficient will be reduced. The actual ranges of various flow patterns over weirs depend on the configuration of the weir. For one typical weir shape the ranges of free overfall, free or submerged hydraulic jump, or subcritical overflow are shown in Figure 9.12.

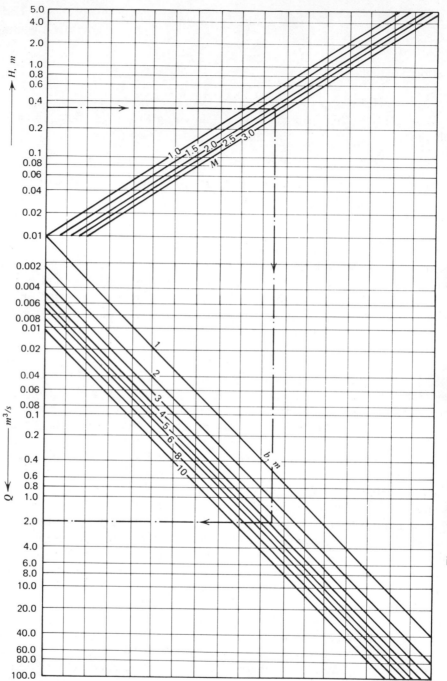

Figure 9.10 Graphical solution for the weir formula with no downstream influence.

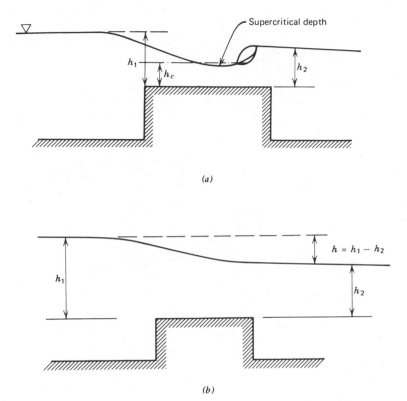

Figure 9.11 Noncritical flow over broad-crested weirs: (a) hydraulic jump over crest; (b) subcritical flow.

Example 9.3

A 4-m (13-ft) wide broad-crested weir is $P = 5.4$ m high above the channel bottom. The normal depth in the channel is 1.2 m when the discharge flowing is 2.0 m³/s (70 ft³/sec). Determine the height at the upstream side if the discharge coefficient is 0.54.

Solution

As P well exceeds the downstream water level there is no downstream influence and therefore Figure 9.10 will be applicable. Computing first the metric weir coefficient M, we get

$$M = c \sqrt{2g} = 0.54(4.43) = 2.4$$

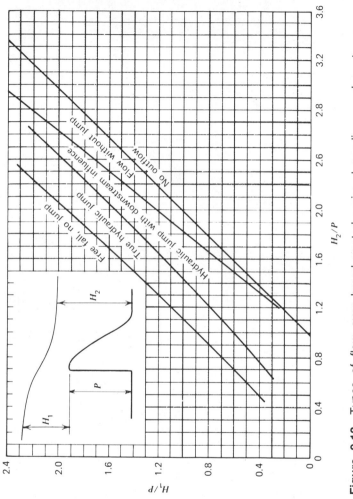

Figure 9.12 Types of flow over broad-crested weirs depending on downstream influence.

Entering Figure 9.10 at $Q = 2$ m³/s and turning at $b = 4$ m width and then at $M = 2.4$, we obtain

$$H = 0.35 \text{ m } (1.1 \text{ ft})$$

9.5 Flow through Culverts

Although simple in appearance, the hydraulic design of culverts is no easy matter. The hydraulic operation of culverts under the various possible discharge conditions presents a somewhat complex problem that cannot be classified either as a flow under pressure or as a free surface flow. The actual conditions involve both of these basic concepts.

The fundamental objective of hydraulic design of culverts is to determine the most economic diameter at which the design discharge is passed without exceeding the allowable headwater elevation. The major components of a culvert are its inlet, the culvert pipe itself, and its outlet with the energy dissipator, if any. Each of these components has a definite discharge capacity. The component having the least capacity will control the hydraulic performance of the whole structure.

One speaks of *inlet control* if, under given circumstances, the discharge of a culvert is dependent only on the headwater above the invert at the entrance, the size of the pipe, and the geometry of the entrance. With the inlet controlling the flow, the slope, length, and roughness of the culvert pipes does not influence the discharge. In this case, the pipe is always only partly full although the headwater may exceed the top of the pipe entrance and hence the flow enters the pipe under pressure.

Short culverts with relatively negligible tailwater elevations almost always operate under inlet control. *Outlet control* occurs when the discharge is dependent on all hydraulic variables of the structure. Figure 9.13 shows the notations relative to these variables. These include the slope S, length L, diameter D, roughness n, tailwater depth h, and headwater depth H. Unless the tailwater level is above the top of the culvert exit, the pipe will be only partly full. This means that the flow in the pipe will be an open channel flow.

Flow in a partially full circular pipe may be analyzed in the manner described in the previous chapter. The depth of the flow relative to the pipe diameter D determines the area of the flow as well as its hydraulic radius. Figure 9.14 aids in the computation of the various hydraulic parameters in relation to those at full flow. From these data the critical depth d_c may be obtained for any discharge. The analytic complexity of this computation is resolved by the graph presented in Figure 9.15. As the critical depth is

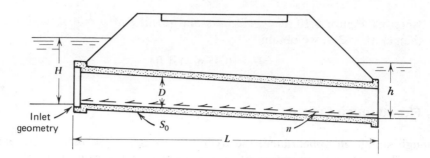

Figure 9.13 Notations in the analysis of culverts.

characterized by the fact that it allows the maximum possible discharge to pass under the prevailing total available energy, it is the essence of desirable culvert operation that the flow be under critical conditions. Equation 8.10 expresses the specific energy in an open channel with respect to the bottom elevation. Rewriting this formula in terms of a partially full circular pipe (in view of the information contained in Fig. 9.14) and introducing two new parameters, namely,

$$a = \frac{\text{actual flow area}}{\text{total area of pipe}}$$

and the discharge factor,

$$q_c = \frac{Q}{D^{2.5} \sqrt{g}} \tag{9.12}$$

the specific energy equation for a partially full pipe may be expressed as

$$\frac{H_0}{D} = 0.81 \left(\frac{q_c}{a} \right)^2 + \frac{y}{D} \tag{9.13}$$

Figure 9.14 shows that the maximum discharge in partially full flow conditions occurs at a depth of $y = 0.93D$. Evaluating the specific energy equation at this depth results in

$$Q_{\text{crit}} = 0.93 \sqrt{g} \, D^{2.5} \tag{9.14}$$

which is the optimum discharge that may be carried by a culvert. Substituting this result into the Chézy–Manning equation (Eq. 8.8) will give the *optimum slope* of a culvert in terms of its roughness and diameter as

$$S_{\text{optimum}} = 75 \frac{n^2}{3 \sqrt{D}} \tag{9.15}$$

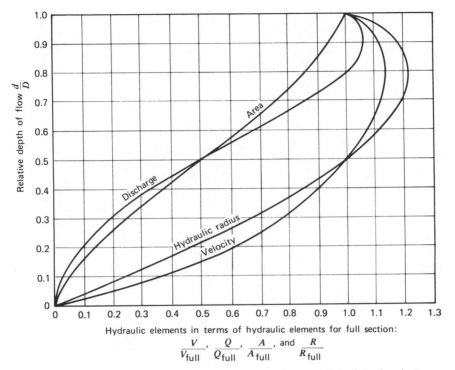

Figure 9.14 Hydraulic parameters Q, v, A, and R in a partially full pipe in terms of those of a full section.

The most common materials used for culverts are concrete and corrugated steel. The roughness in both cases is usually assumed to be constant for any flow depth. Common values used in practice are $n = 0.012$ for concrete pipe and $n = 0.024$ for corrugated steel pipe. Although the selection of culvert materials is made in practice on a competitive economic basis, the fact that the roughness of corrugated steel pipes is twice that of concrete pipes suggests that their selection be based on hydraulic considerations also. On hilly terrains where the culvert slope is expected to be relatively steep and the flow through the culvert gains considerable energy, corrugated steel pipes offer energy dissipating advantages. On flat terrains energy loss through a culvert is undesirable; hence, concrete pipes are more suitable.

Evaluating Equation 9.15 for concrete as well as for corrugated metal indicates that the optimum slope of a concrete culvert is about 0.01, while for corrugated metal pipe it ranges between 0.06 for smaller pipes and 0.04 for larger pipes.

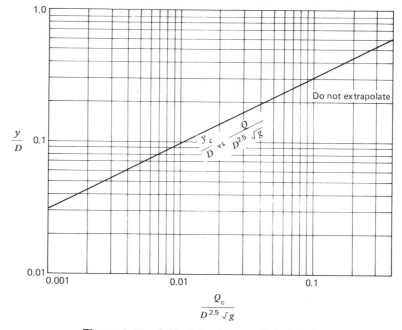

Figure 9.15 Critical depth in partially full pipe.

There is a great variety of design charts, nomograms, specifications, and design recommendations for culverts. Most state and federal organizations have produced extensive literature for their systematic design. Manufacturers of concrete and metal culverts are another source of design information. The typical design procedure is to first determine the expected discharge and the allowable headwater elevation. By Equation 9.14 a trial pipe diameter may be selected. By determining the material to be used, the slope of the culvert may be computed by Equation 9.15. This will allow the determination of the length and the tailwater elevation. With all this information provided, the next step is to define whether there is outlet or inlet control. For both cases design nomographs are generally available. Actually there is little need in practice to define which type of control is present. The design computation may be carried out for both cases and the result giving the more critical answer will be the controlling one. There are two design nomographs included here. Figure 9.16 is for short pipes operating under inlet control. Figure 9.17 is for long concrete pipes operating under outlet control, flowing full. The k coefficients shown in the latter nomograph are entrace loss coefficients. Selected values for these entrance loss coefficients are listed in Table 9.1.

Table 9.1
Culvert Entrance Loss Coefficients

Type of entrance	Entrance loss coefficient, k
Headwall entrances	
Rounded edge ($R = 0.25D$)	0.10
Rounded edge ($R = 0.15D$)	0.15
Grooved edge	0.19
Square edge (metal pipe)	0.43
Projecting entrances	
Grooved thick wall	0.25
Squared edge, thick wall	0.46
Sharp edge (metal pipe)	0.92

Example 9.4

A 1.5 m (4.9 ft) diameter concrete culvert shown in Figure E9.4 is 30 m (98.4 ft) long and its entrance has a grooved edge in a headwall ($k = 0.2$). Determine the optimum discharge, the optimum slope, and the corresponding velocity.

Solution

By Equation 9.14 the optimum discharge is

$$Q_{\text{crit}} = 0.93 g^{1/2} D^{2.5} = 8.02 \text{ m}^3/\text{s} \ (283 \text{ ft}^3/\text{sec})$$

By Equation 9.15 the optimum slope is

$$S_{\text{optimum}} = 75n/D^{0.33} = 0.01$$

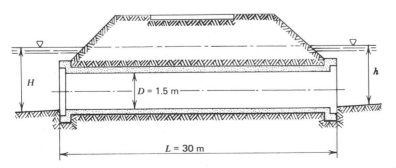

Figure E9.4

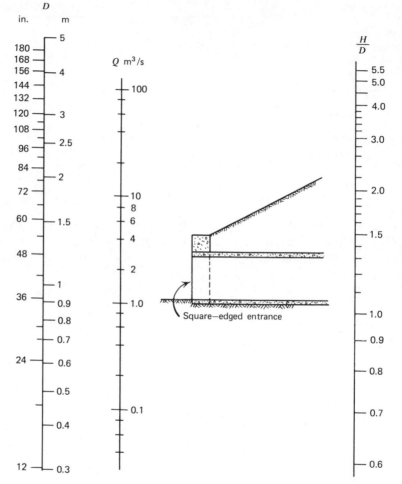

Figure 9.16 Nomographic solution for a typical culvert under inlet control. Square edged entrance.

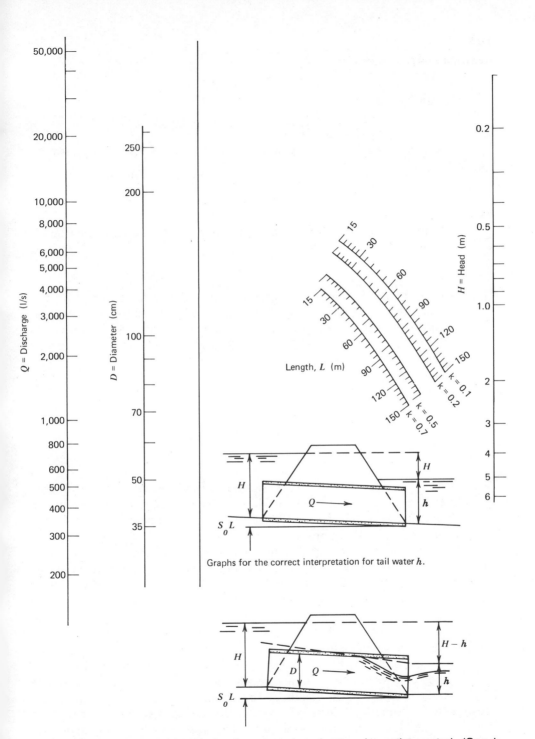

Figure 9.17 Nomographic solution for concrete culvert under outlet control. (See *k* coefficients in Table 9.1.)

193

The area of the pipe is

$$A = \pi D^2 / 4 = 1.76 \text{ m}^2 \, (18.94 \text{ ft}^2)$$

Using Figure 9.14 the velocity at $d/D = 0.93$ is $v = 1.1(v_{full})$. Therefore,

$$v = 1.1Q/A = 1.1(8.02)/1.76 = 5 \text{ m/s} \, (16.4 \text{ ft/sec})$$

Example 9.5

Determine the head loss at optimum discharge in the culvert described in Example 9.4.

Solution

With optimum discharge it is assumed that the water flows at critical velocity throughout the pipe, filling it to 93 percent of its diameter. Accordingly, the pipe roughness will influence the flow and the culvert is under outlet control. Using Figure 9.17, the head loss across the culvert will be

$$H_L = 1.41 \text{ m} \, (4.6 \text{ ft})$$

9.6 Design of Sewers

There are three main types of sewers: sanitary sewers, which carry household or industrial sewage; storm sewers, which collect rainwater in urban areas; and combined sewers, which do both, but are not recommended anymore due to environmental considerations.

Most sewers, particularly smaller ones, are circular in cross section. Other shapes are also used; these generally provide the efficient conveyance of low flows by a small curvature on the bottom, and large cross sections on their higher parts for carrying larger discharges. The minimum recommended sanitary sewer diameter is 20 cm (8 in.), although in some cases a 15 cm diameter is also permitted. For storm sewers a 45-cm diameter minimum size is recommended, although for short runs 30 cm may be allowed.

The horizontal alignment of sewers is controlled by their purpose and by maintenance considerations. Sanitary sewers are usually laid along the middle of the street to allow equal distances for connecting houses on both sides. Storm sewers, usually off-center, in most cases are located along the edge of the pavement. Sewers are composed of straight sections connecting manholes that are rarely farther apart than 100 or 150 meters. This permits relatively easy cleaning in case of blockage.

Vertical alignment of sewers is guided by a number of rules that on occasion may be broken. Sanitary sewers are to be placed deep enough to permit gravity drainage from all basements of residences served. In exceptionally deep positions some basements may have to be provided with pumps that raise the sewage to the level of the main sewer, as the extra expense of excavating deeper for the main sewer might not be warranted. The depth of the sewer should include consideration for the need for turns and the slope of the sewage line leading out of basements. Attempts should be made to follow the slope of the ground surface with the slope of the sewer, which will minimize the depth of the excavation required. When selecting the slope of the sewer as a part of the hydraulic design, the velocity at full flow should be kept within 0.5 and 3 m/s. The smaller velocities will lead to sedimentation and clogging; the higher velocities will cause erosion of the pipe walls. The designer should avoid the temptation to use a pipe larger than necessary to carry the design discharge. Larger pipes would allow flatter slopes, hence less excavation, but the resultant lower velocities would lead to clogging in the pipe, particularly between high flows. Changes of pipe diameter should always occur at manholes. The top of the two pipes are usually placed at the same elevation, or the exiting pipe is located below that. Contents of a larger pipe should never discharge into a smaller one as it always leads to blocking.

The required discharge capacity of sanitary sewers is dependent on the drainage area served and its population density. A reasonable value is 400 liters per person per day. More specific information is available in textbooks on sanitary engineering and in local codes. In addition to the sewage carried, the sanitary sewers are always subjected to groundwater infiltration, regardless of the degree of care by which they are constructed. Storm sewers' discharge capacity is dependent on the individual drainage area of each manhole and storm inlet. The discharge for each of these drainage areas is determined by the methods discussed in Chapter 1. For smaller projects the Rational Method furnishes acceptable results. In addition to the resulting discharge the time of concentration must also be determined, that is, the times at which the peak runoff will reach the storm sewer at each manhole and the main trunk sewer at the junction of each tributary. To determine the magnitude of the peak flow at the various portions of the sewer line along the way, the time of travel of each pipe length must be determined by dividing the length by the velocity of the flow. These, when added to the time of concentration on the surface, will indicate how the peaks from the various tributaries will add up.

The hydraulic design of sewerage systems is relatively simple. The basic prinicple is to design the pipe as an open channel flowing just full. To do this we have to use the concept of normal flow, in which the hydraulic gradient is the slope of the top of the sewer pipe. Knowing the available sizes and roughness of commercial pipe, the designer can determine the velocity and the corresponding slope for each pipe section. As long as the velocity is within the allowable limits, and all other fundamental design rules are met, the pipe may be drawn with the slope computed. The design is usually carried on together with the drawing of the profile of the sewer.

Example 9.6

Determine the discharge capacity of a 20 cm (8 in.) diameter concrete sewer pipe laid at a slope of 0.06.

Solution

Using the Chézy–Manning equation (Eq. 8.8 and Fig. 8.4) with $n = 0.012$ and $R = D/4$, we find that the velocity of the flow is

$$v = 2.8 \text{ m/s}$$

The cross-sectional area of the pipe is

$$A = \pi D^2/4 = 0.031 \text{ m}^2$$

Therefore the discharge in the sewer is

$$Q = v \cdot A = 2.8(0.031) = 0.088 \text{ m}^3/\text{s} \ (3.1 \text{ ft}^3/\text{sec})$$

Problems

9.1 A 1.2 m diameter smooth steel pipe is used as a bottom drain in a small reservoir. The length of the pipe is 18 m. The pipe has a rounded entrance. Determine the flow through the pipe if the height of the water in the reservoir is 3.4 m over the centerline of the pipe. (*Ans.* 8.2 m³/s)

9.2 The downstream water level of a horizontal channel is 2 m above the bottom. The discharge per unit width is 1.2 m³/s/m. Determine the conjugate depth on the upstream side of a corresponding hydraulic jump. (*Ans.* 0.14 m)

9.3 The Froude number of the upstream side of a hydraulic jump is 4.0. Determine the upstream water level if the downstream depth is 3.8 m. (*Ans.* 0.73 m)

9.4 A 3-m wide gate is opened to a 0.6 m height. The depth above the gate is 5 m; below the gate it is 2 m. Determine the type of flow and discharge under the gate. (*Ans.* 12 m³/s)

9.5 The discharge coefficient of a broad-crested weir is 0.5. The width of the weir is 5 m. The height of the water in the reservoir is 1.3 m above the crest of the weir. The downstream water level is below the crest elevation. Determine the discharge flowing over the weir. (*Ans.* 16.4 m^3/s)

9.6 A 40-m long culvert is 1.5 m in diameter. The depth of water upstream should not exceed 2.0 m. The culvert is to be made of concrete. Determine the discharge under normal inflow. (*Ans.* 3.5 m^3/s)

Chapter 10
Seepage

10.1 Basic Concepts

The flow of water through soils follows the same principles of hydraulics as it does when flowing through closed or open conduits. Sand, silt, clay, and many sedimentary rocks have interconnected open pores between their solid particles. Through these pores water may flow either under pressure or by gravity. As the tiny channels between the grains are random in their size and orientation, the direction and magnitude of the *seepage velocity* within the soil is random as well. If, however, the porosity of the soil is known, then the *discharge velocity* of the flow can be determined by taking the ratio of the discharge flowing through a certain cross section and the area of the cross section of the soil. The porosity η of a soil is interpreted as the ratio of the volume of voids to the whole volume of the soil. Table 10.1 shows some typical porosity values for selected natural soils. On the average the percentage of open spaces in a cross section of the soil is the same as its porosity determined volumetrically. Therefore the relationship between seepage velocity and the other parameters is

$$v_{\text{seepage}} = \frac{v_{\text{discharge}}}{\eta} = \frac{Q}{\eta \cdot A} \qquad (10.1)$$

Table 10.1
Typical Porosity Values of Soils

Soil type	Porosity, η
Uniform loose sand	0.46
Uniform dense sand	0.34
Glacial till	0.20
Glacial clay, soft	0.55
Glacial clay, stiff	0.37
Soft very organic clay	0.75

in which Q is the discharge through the soil surface A. In the remainder of this chapter the term velocity will mean discharge velocity unless stated otherwise.

Seepage through porous soils is analyzed by the equation of continuity and by the Bernoulli equation. To apply these, consider the case of a soil tube shown in Figure 10.1. When the velocity of the flow is tangential to the surface of the soil tube at all points, then the discharge flowing into the tube at one end is the same as the discharge flowing out at the other. Bernoulli's equation may be written between the two ends resulting in

$$\frac{v_1^2}{2g} + \frac{p_1}{\gamma} + z_1 = \frac{v_2^2}{2g} + \frac{p_2}{\gamma} + z_2 + h_L \qquad (10.2)$$

in which h_L is the energy loss between the two points. This energy loss takes place due to the viscous shear between the water and the soil particles.

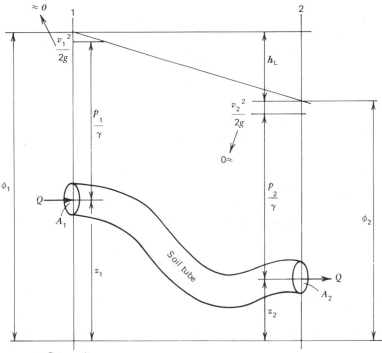

Figure 10.1 Seepage through a soil tube. Notations for the energy equation (Eq. 10.2).

In almost all cases of groundwater flows the velocity of the seepage is very small. As the square of a very small number is even smaller, the kinetic energy terms in our energy equation are negligible. A further simplification of Equation 10.2 comes about by introducing the term called *potential,* which is defined as the sum of the pressure and elevation energies, that is,

$$\phi = p / \gamma + z \qquad (10.3)$$

By dropping the kinetic energy terms and introducing potential notation, Equation 10.2 simplifies into

$$\phi_1 - \phi_2 = h_L \qquad (10.4)$$

When imagining this phenomenon it is worthwhile to note that the viscous shear loss h_L is not the result of heat dissipating into the atmosphere, like in pipes and open channels, but of a drag force acting on the soil matrix. If the potential difference is large enough, the seeping fluid may pick up and carry away some of the unsupported particles from the soil mass. Even larger potential differences lead to the so-called "quick condition," also known in chemical engineering as "fluidized bed." There the drag forces exceed interstitial forces caused by gravity and other factors causing the soil to lose its strength and behave as a liquid. As the gravity forces act vertically downward, quick condition in a soil requires upward flow.

Applying Equation 10.4 to a very small soil element of ΔL length and denoting the energy loss as Δh, one gets

$$\frac{\Delta \phi}{\Delta L} = \frac{\Delta h}{\Delta L} = i \qquad (10.5)$$

in which i is referred to as the *energy gradient* at the point in the soil mass where the small element is considered. Since the ΔL length is a directional quantity, the energy gradient possesses both direction and magnitude.

Scientific analysis of seepage problems is rooted in *Darcy's equation,* which states that the velocity of the flow is proportional to the energy gradient,

$$v = k \cdot i = k \, \frac{\Delta \phi}{\Delta L} \qquad (10.6)$$

in which the constant of proportionality is Darcy's permeability coefficient. Values for Darcy's k for typical soils are shown in Table 10.2. The permeability coefficient of various soil layers is usually determined by pumping tests in drilled wells, laboratory tests on soil samples from bore holes, or in an indirect and approximate manner by sieve analysis of the soil samples. Since neither of these methods is free of pitfalls, reliable information on the permeability coefficient in actual practice is rare.

Table 10.2
Darcy's Permeability Coefficient k

Soil type	Average grain size (mm)	Range of k coefficient (cm/s)	Order of magnitude of k (cm/s)
Clean gravel	4–7	2.5–4.0	1
Fine gravel	2–4	1.0–3.5	1
Coarse, clean sand	0.5	0.01–1.0	10^{-1}
Mixed sand	0.1–0.3	0.005–0.01	10^{-2}
Fine sand	0.1	0.001–0.05	10^{-3}
Silty sand	0.02–0.1	0.0001–0.002	10^{-4}
Silt	0.002–0.02	0.00001–0.0005	10^{-5}
Clay	0.002	$10^{-9}-10^{-6}$	10^{-7}

10.2 Flow Nets—Confined Flow

The first step in the analysis of a seepage problem is the determination of the shape of the flow field. As three-dimensional seepage problems are generally much more difficult to handle, here the assumption will be made that all seepage problems considered are "two dimensional," that is, the shape of the flow field is identical regardless of the location where the permeable structure is bisected. Examples for such two-dimensional seepage problems are: seepage through earth dams and levees, seepage under long and straight sheetpiles and concrete dams, and seepage into long trenches and galleries. For the sake of clarifying the basic concepts, the arrangement shown in Figure 10.2 will be considered. In the sand-filled tank depicted in this figure the region enclosed by points A through K is the flow field. Flow fields that are geometrically defined to begin with are called *confined flow* fields. Another case of seepage through a porous medium is shown in Figure 10.3. Here the flow field is enclosed by points A through E. The boundary between A and B is exposed to the atmosphere and is called the *free surface*. Between B and C the flowing water exits the flow region. This boundary portion is called the *surface of seepage*. At the beginning of the analysis of such problems the location of point B and the shape of line $A–B$ are unknown. Hence the flow field is initially undefined. Flows of this kind are called *unconfined flows*.

In Figure 10.2 one may observe that the values of ϕ (see Eq. 10.4) are defined between points A and K as to be equal,

$$\phi_{A-K} = M + H_1 = \phi_{\text{inflow}} \qquad (10.7)$$

Likewise the potential between points F and E is also a constant value equal to

$$\phi_{F-E} = M + H_2 = \phi_{\text{outflow}} \qquad (10.8)$$

The difference between these inflow and outflow boundaries of our flow region is the potential difference of

$$\phi_{\text{inflow}} - \phi_{\text{outflow}} = h_L = S \qquad (10.9)$$

which is the external energy that drives the flow through the flow field. In case of a flow field that is symmetric around a vertical line, as in Figure 10.2 through points G and C, one may further realize that half of the total energy loss had occurred in the first half of the flow region; therefore the value of the potential along this line is constant and equal to

$$\phi_{G-C} = S / 2 \qquad (10.10)$$

The term S in seepage computations is often called the *drawdown*. Regardless of the actual magnitude of the inflow and outflow potentials, the driving energy depends only on their difference, the magnitude of S.

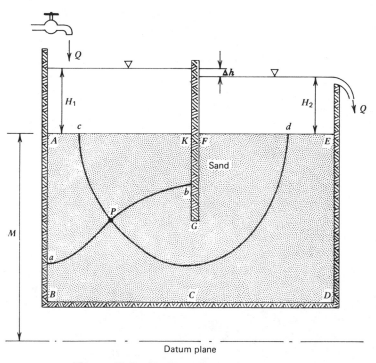

Figure 10.2   Confined flow in a sand tank.

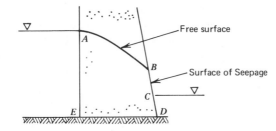

Figure 10.3 Unconfined flow through a porous wall.

The lines between *A* and *B*, *F* and *E*, as well as *G* and *C* (if there is symmetry) are distinguished by having a constant potential value along them. Any line within a flow field on which the value of the hydraulic potential ϕ is constant is called *equipotential line*. For each point of a flow field there exists a unique potential value, since both the elevation and the hydrostatic pressure that exists there can be uniquely defined. Through each point within a flow field crosses one equipotential line. Usually we single out some primary equipotential lines representing a certain percentage or fraction of *S*. In solving seepage problems we seek the location of equipotential lines such as $S/4$, $2S/4$, $3S/4$, or $0.2S$, $0.4S$, $0.6S$, and $0.8S$. In the case of point *P* shown in Figure 10.2, the equipotential line is sketched between *a* and *b*. In the beginning neither the value nor the location of this line is known.

In the flow field the total discharge flowing is denoted by *Q*. This discharge is flowing between the two flow boundaries defined by the line through *A–B–C–D–E* on the one hand and the line *K–G–F* on the other. Being boundaries to our "flow tube" (along with two parallel vertical planes a unit distance apart), it is postulated that the velocity of the flow along these planes is tangential. The two such boundaries described above are called *limiting stream lines*.

For each point within the flow field there exists a velocity vector that indicates the magnitude and direction of the flow at that point. Equation 10.6 defines this velocity in terms of the function ϕ. Because flow lines within the flow field do not cross each other, one may realize that this velocity vector separates the flow through the flow field such that on one side of it flows a portion of *Q* while on its other side flows the remainder. In fact, there is a line connecting adjacent points in the flow field such that the ratio of *Q* flowing on each side is constant. Velocity vectors at all of these points are tangent to this line. When the flow is steady, such lines are called *stream lines*. Stream lines are commonly denoted by the Greek letter ψ (psi). Because a constant portion of *Q* passes between two stream lines, one may write

$$\psi_1 - \psi_2 = \Delta\psi = \Delta Q \tag{10.11}$$

Primary stream lines within a flow field are selected such that they represent certain percentages or fractions of Q such as $0.25Q$, $0.5Q$, $0.75Q$, and so on. In the case of point P shown in Figure 10.2 the stream line in question extends from c to d. In the beginning neither the value nor the location of this stream line is known. In the analysis of a seepage problem the solution is aimed to determine the location of the primary equipotential and stream lines. Once these are known, specific questions relating to the seepage pressures, the discharge Q, and the velocity distribution may be answered without difficulty. There are several major ways a solution can be found. These include: the analytical solution known as *conformal transformation; numerical methods* using digital computers; experimental methods based on *electric analogy;* and graphical construction of *flow nets.* The latter will be introduced here. Other methods of solution are considerably more complex than the graphical solution.

The basic concept on which the solution of two-dimensional flow-net problems is based is that the network of primary equipotential and stream lines are mutually perpendicular to each other at all points of the flow field. Since the water seeks the path of least resistance, the direction of the velocity at all points, such as point P in Figure 10.2, must be perpendicular to the equipotential line. Hence if one manages to construct two sets of lines such that they will intersect each other perpendicularly, including the limiting equipotential and stream lines that are known in the beginning, then the result is the solution sought.

In considering further the solution of a flow net, let us rewrite Equation 10.6 in a generalized manner. Substituting Q/A in place of v, writing S in place of $d\phi$ to express the energy loss across the whole flow field, introducing the length L for an average (statistical mean) stream line length, and then rearranging, one gets

$$\frac{Q}{kS} = \frac{A}{L} = F \tag{10.12}$$

in which the left-hand side is called the *relative discharge* and F on the right-hand side is called the *form factor.* The relative discharge is a meaningful dimensionless term representing the flow through a unit depth of a two-dimensional flow field. Equation 10.12 indicates that this relative discharge depends only on the geometry. A in this equation is an "average" cross-sectional area, measured as the statistical mean length of the equipotential lines, while L, as stated above, is the statistical mean length of the stream lines. While determining the mean lengths of these initially unknown quantities would present major difficulties, their ratio may be obtained with relative ease. To obtain the form factor one has to determine only the ratio of the two statistical quantities by expressing them in identical scale. This may be done by sketching a flow net in which the two perpendicular families of lines form squares. Of course these will be distorted squares since the equipotential lines and stream lines are curved. Nevertheless, the intersections of the primary stream lines and equipotential lines should be perpendicular and should be drawn such that each square embraces a circle so that it is touched at four points. Figure 10.4 shows the essence of this concept.

The graphical construction of flow nets usually starts by drawing a few main stream lines (or potential lines) within the boundaries of the flow field. These, of course, are only rough approximations. Following this the other family of lines is drawn starting at one end of the flow field in such a manner that they form a proper "square" net. One must remember that the boundary lines are also parts of the net. Invariably the first trials result in incorrectly shaped "squares" or imperfect (nonperpendicular) intersections of lines. Hence the graphical process involves frequent erasures in parts of the flow net until an approximately correct plot is produced. One will soon learn that even small local changes may cause significant variations in the whole plot. One of the most frequent mistakes made by beginners is the attempt to make the net too fine, drawing too many small squares. Starting with more than two or three stream lines in the beginning leads to this problem. A convenient method is to draw the fixed flow boundaries on an 8½ × 11 in. paper, which is then inserted in a transparent plastic cover (like Copco PS-5, for example).

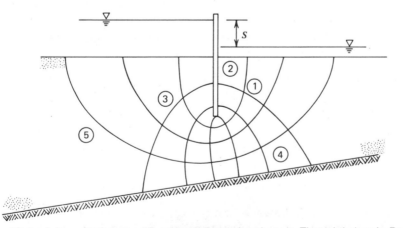

Figure 10.4 A flow net for a confined two-dimensional case. First trial sketch: Points 1 and 2—correct intersection, proper square; point 3—obviously incorrect intersection; point 4—incorrect, elongated square; point 5—major error, squares are too long, left side to be redone entirely.

The trial flow net is drawn on the plastic sheet where it can be erased and corrected with ease.

Once a reasonably appearing flow net is drawn, the form factor may be determined by

$$F \text{ (form factor)} = \frac{\text{number of stream lines}}{\text{number of equipotential lines}} \qquad (10.13)$$

In both numbers the first initial stream or equipotential line is not counted. The result may then be substituted into Equation 10.12 from which the discharge, or the unknown energy loss S, may be expressed. The flow net can be used for the determination of hydrostatic pressures of certain points by the use of Equation 10.3.

One of the most important uses of the flow net is the evaluation of stability along the exit surface with respect to quick condition. For common granular soils the velocity at which the soil will be fluidized is approximately equal to the permeability coefficient of the soil. In light of Equation 10.6 this occurs when

$$\frac{v}{k} = \left(\frac{d\phi}{dL} \right)_{\text{crit}} = 1 \qquad (10.14)$$

In other words, the critical exit gradient equals unity. In flow regions such as depicted in Figure 10.2 this critical condition occurs where the stream line is the shortest, namely at point F. With the flow net available the approximate value of the exit gradient at F can be determined by evaluating the energy drop across the last square of the flow net and dividing the result by the elevation difference between the last and the immediately preceding potential line.

Example 10.1

Determine discharge and the uplift pressures along the horizontal base of the dam shown in Figure E10.1. The coefficient of permeability is 10^{-4} cm/s.

Solution

First a flow net is drawn within the zone of seepage as shown in Figure 10.1. The equipotential and stream lines are numbered in the drawing. With the form factor being 0.33, the relative discharge by Equation 10.12 is

$$Q/kS = 0.33$$

where S is 15 m and k is 10^{-4} cm/s. Hence the discharge for a unit width is

$$Q = 0.33(15)10^{-6} = 5 \times 10^{-6} \text{ m}^3/\text{s (18 liters/hr)}$$

Next we consider the potential lines along the horizontal base of the dam.

The potential energy along each of these lines is defined as being constant and equal to the total potential difference across the whole flow field times the serial number of the line in question divided by the total number of potential lines in the flow net. Therefore, the value of the potential at the intersection of, say, line number 10, is

$$h_{10} = H \times \frac{10}{15} = 10 \text{ m}$$

since $H = 15$ m. By Equation 10.3 the term *potential* is defined as the sum of the elevation and pressure energies existing at the point. To define the elevation energy we first pick a datum plane. In our case the base line itself is a convenient datum, making the y zero. Hence, in this case

$$h = p/\gamma$$

Thus the unbalanced uplift pressure due to the percolating water below the base of the dam equals

$$p_{10} = 10 \times 9810 = 98,100 \text{ N/m}^2 \text{ (2060 lb/ft}^2)$$

At all other intersections the pressures may be computed likewise. To obtain pressures at intermediate points the squares of the flow net may

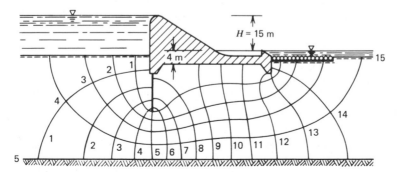

Figure E10.1

be further subdivided (like at $h = 9.5$, etc.). To obtain the pressure diagram on the base the individual points are connected by a smooth line.

10.3   Flow Nets—Unconfined Flows

In Figure 10.3 an example is shown for the case when the flow field is not initially determined. The free surface may rise to any level determined by the flow underneath it. However unknown its position may be, it shall follow the basic physical law of hydraulics—the theorem of minimum energy. Accordingly there is one single solution for such problems: the one at which the most water flows through at the optimum utilization of the energy available for the flow. Once the location of a free surface is known, the problem becomes a confined flow problem and can be handled according to the concepts outlined in Section 10.2.

In determining the position of the free surface two fundamental definitions are utilized:

a.   The free surface is a limiting stream line, and, as such, all potential lines are perpendicular to it.
b.   The pressure along the free surface is atmospheric, hence the value of all potential lines on the free surface equal its elevation at their point of intersection (note Eq. 10.3 with $p = 0$).

With the above concepts kept in mind the graphical solution of unconfined flow problems can be obtained by the following trial and error approach:

First, a trial shape is drawn for the free surface making sure that it is started from the entrance in a direction perpendicular to the initial potential line, and that it is falling gradually in the direction of the flow. To sketch a trial free surface one of its points must be known initially. In Figure 10.3 point A at the entrance is definitely known. But if the flow field shown in Figure 10.4 would be altered so that the groundwater would not rise to the surface beyond the sheetpile but continue to flow under ground level, no point of the free surface would be known and the initial trial surface could not be drawn until one of its points would be determined by drilling a bore hole.

After this trial surface is drawn a flow net is constructed for the now confined flow field. Care must be exercised that the equipotential lines are joining the free surface in a perpendicular manner.

Once the flow net is drawn in a satisfactory manner, the elevation of each primary potential line is determined at the point where it joins the free surface. These elevations must equal the value of the equipotential lines or else the trial free surface was drawn incorrectly. Figure 10.5 will aid in the interpretation of this procedure.

If an error is indicated in the alignment of the free surface, it is to be redrawn. In Figure 10.5 the trial surface is apparently slightly lower than it should be. Once a new trial surface is established, the procedure is repeated. Experience shows that with some practice one rarely needs more than two trials before a satisfactory solution is obtained.

10.4 Discharge From Wells

Formulas to determine the discharge of wells were first developed by J. Dupuit more than a century ago. Wells extract water from permeable water-bearing layers called *aquifers*. One may distinguish between shallow wells that usually extract water from layers characterized by a free surface, and deep wells taking water from a buried aquifer in which the water is under hydrostatic pressure. The latter are called artesian aquifers.

Shallow wells may be either vertical that are dug, driven, or drilled, or horizontal, such as *galleries* or *collector pipes* radially arranged around a vertical concrete shaft.

With certain theoretical simplifications Dupuit derived the discharge formula of a *shallow vertical well* under steady flow conditions, extracting water

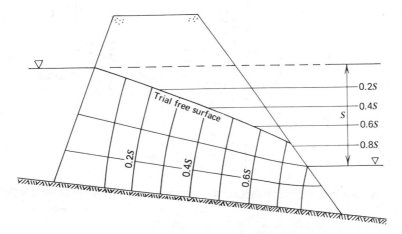

Figure 10.5 Graphical solution of unconfined seepage through a permeable dam.

from a horizontal water-bearing layer of infinite extent. His formula is commonly written as

$$Q = \frac{k\pi(H^2 - h^2)}{\ln(R/r)} \tag{10.15}$$

in which the variables are as shown in Figure 10.6. The term R in Equation 10.15 is called the "radius of influence," an empirically determined factor that represents the distance from the well at which the free surface height remains unchanged (H) while the well is being pumped. One of the empirical equations known in the literature for the radius of influence is Sichardt's, which expresses R as a function of the drawdown in the well, S, and the permeability of the soil, in the form of

$$R = 3000S\sqrt{k} \tag{10.16}$$

where k is to be substituted in meters per second. This formula is dimensionally incorrect and may be used for rough estimates only.

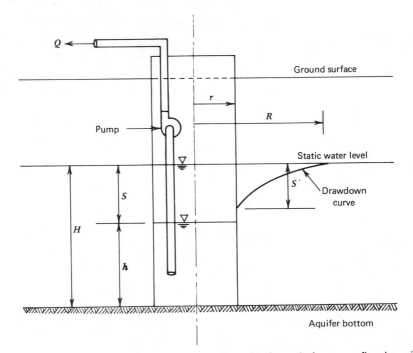

Figure 10.6 Notations for Equation 10.15, Dupuit's formula for unconfined aquifers.

Artesian wells obtain their water from one or more water-bearing layers located at greater depths in which the water is under hydrostatic pressure. For the discharge of a horizontal artesian layer of T thickness, Dupuit's derivation yields

$$\frac{Q}{kTS} = \frac{2\pi}{\ln(R/r)} \qquad (10.17)$$

where R, again, is the radius of influence given by Equation 10.16.

Example 10.2

An artesian well tapped a 12-m (39.36-ft) thick layer of silty sand at a depth of 500 m. The static water level of the aquifer is 13 m (42.64 ft) above the ground in a closed standpipe. Sieve analysis indicated that the average grain size of the soil is 0.08 mm (0.003 in.). The well diameter is 0.02 m (0.79 in.). How much discharge is expected if the water is allowed to flow out at a level 2 m (6.56 ft) above ground?

Solution

By Table 10.2 the approximate value of the permeability coefficient is taken as

$$k = 10^{-4} \text{ cm/s} = 10^{-6} \text{ m/s}$$

The drawdown is $13 - 2 = 11$ m. Using Sichardt's equation (10.16) to obtain the radius of influence, one gets

$$R = 3000(11)10^{-6/2} = 33\text{m}$$

Substituting into the discharge formula for artesian wells (Eq. 10.17),

$$Q = kTS \ (2(3.14))/\ln(33/0.01)$$

$$= 10^{-6}(12) \ 11(2) \ 3.14/8.1$$

$$= 10.2 \times 10^{-5} \text{ m}^3/\text{s} = 3672 \text{ liters/hr}$$

Problems

10.1   Differentiate between discharge velocity and seepage velocity.

10.2   State and explain the definition of the term "potential."

10.3   What is "energy gradient"?

10.4 Explain Darcy's equation.

10.5 What is the difference between confined and unconfined seepage?

10.6 Define limiting stream lines and limiting potential lines.

10.7 What is relative discharge in seepage?

10.8 How is the form factor determined for a confined flow field?

10.9 How does one compute the hydrostatic pressure at a point in a flow field?

10.10 Why are flow nets drawn such that the intersecting sets of curves form a square net?

10.11 Why is the velocity vector perpendicular to the equipotential line?

10.12 Describe the procedure used in determining the correct position and shape of a free surface.

10.13 What two requirements are satisfied on the free surface?

10.14 At what point will "quick condition" occur?

10.15 Differentiate between unconfined and artesian aquifers.

Appendix 1
Conversion Factors for Volume, Discharge, and Rainfall

Volume Conversion Factors

	Cubic Inch	U.S. Gallon	Imperial Gallon	Cubic Foot	Cubic Yard	Cubic Meter	Acre-Foot	Second-Foot-Day
Cubic inch	1	0.00433	0.00361	5.79×10^{-4}	2.14×10^{-5}	1.64×10^{-5}	1.33×10^{-8}	6.70×10^{-9}
U.S. gallon	231	1	0.833	0.134	0.00495	0.00379	3.07×10^{-6}	1.55×10^{-6}
Imperial gallon	277	1.20	1	0.161	0.00595	0.00455	3.68×10^{-6}	1.86×10^{-6}
Cubic foot	1728	7.48	6.23	1	0.00370	0.0283	2.30×10^{-5}	1.16×10^{-5}
Cubic yard	46,656	202	168	27	1	0.765	6.20×10^{-4}	3.12×10^{-4}
Cubic meter	61,000	264	220	35.3	1.31	1	8.11×10^{-4}	4.09×10^{-4}
Acre-foot	7.53×10^{7}	3.26×10^{5}	2.71×10^{5}	43,560	1610	1230	1	0.504
Second-foot-day	1.49×10^{8}	6.46×10^{5}	5.38×10^{5}	86,400	3200	2450	1.98	1

Discharge Conversion Factors

	Million Gallons/Day	Gallons/Minute	Cubic Foot/Second	Cubic Foot/Day	Acre-Foot/Day	Meter^3/Second	Meter^3/Hour	Meter^3/Day	Millimeter/Year/Kilometer^2	Inch/Year/Mile^2
Million gallons/day	1	694	1.55	134×10^3	3.07	43.8×10^{-3}	157.68	3784	1381	20.98
Gallons/minute	1.44×10^{-3}	1	2.227×10^{-3}	193	4.42×10^{-3}	6.31×10^{-5}	0.227	5.448	1.988	0.03
Cubic foot/second	0.646	449	1	86,400	1.98	0.0283	101.88	2445	892.4	13.56
Cubic foot/day	7.48×10^{-6}	5.19×10^{-3}	1.16×10^{-5}	1	2.3×10^{-5}	3.28×10^{-7}	1.18×10^{-3}	0.028	0.0120	1.55×10^{-4}
Acre-foot/day	0.326	226	0.504	43,560	1	0.0143	51.48	1235	450	6.84
Meter^3/second	22.8	15,800	35.3	3.05×10^6	70	1	3600	86,400	31,536	479
Meter^3/hour	6.34×10^{-3}	4.40	9.81×10^{-3}	847.5	0.0194	2.77×10^{-4}	1	24	8.76	0.133
Meter^3/day	2.64×10^{-4}	0.183	4.09×10^{-4}	35.7	8.1×10^{-4}	1.157×10^{-5}	41.6×10^{-3}	1	0.365	5.55×10^{-3}
Millimeter/year/kilometer^2	7.24×10^{-4}	0.502	1.12×10^{-3}	98	2.22×10^{-3}	3.17×10^{-5}	0.114	2.74	1	0.0152
Inch/year/mile^2	0.047	33.33	0.074	6451	0.146	2.1×10^{-3}	7.53	180	65.79	1

Rainfall and Runoff Conversions

	Millimeter/Year	Inch/Year	Liter/Second/Kilometer^2	Meter^3/Year/Kilometer^2	Foot^3/Second/Mile^2
Millimeter/year	1	0.0394	0.03169	1000	0.3468
Inch/year	25.4	1	0.805	25,400	8.85
Liter/second/kilometer^2	31.55	1.1242	1	31,550	10.94
Meter^3/year/kilometer^2	0.001	0.392×10^{-4}	0.316×10^{-4}	1	3.47×10^{-4}
Foot^3/second/mile^2	2.88	0.113	0.0914	2880	1

Appendix 2
Laws of Hydraulic Modeling

Hydraulic models of a full scale "prototype" are designed such that the motion and forces in them are similar to those in the prototype. To attain this the designer should be familiar with the concept and laws of geometric, kinematic, and dynamic similarity.

Geometric similarity means that all corresponding lengths in the model and prototype are of the same ratio. Geometric scale of a model is selected so that it can be built and operated conveniently in a hydraulic laboratory and still be big enough to measure velocities in it with sufficient accuracy. Model scales of spillways, conduits, and similar structures range between 1:15 and 1:50. Models of rivers and harbors range from 1:100 to as small as 1:2000. With such small scales water depths in the model would be so small that the effects of surface tension and roughness could be overbearing. In such cases the vertical scale may differ from the horizontal scale; we would then speak of distorted models.

To satisfy the fundamental requirement that the model is to operate in a manner similar to the prototype, one has to recognize that in each the flow follows Newton's law of motion. For the two systems,

$$F_p = m_p \cdot a_p$$
$$F_m = m_m \cdot a_m \qquad \text{(A.1)}$$

where subscripts p and m represent prototype and model. The acceleration is the time rate of velocity change

$$a = \Delta V / \Delta T$$

and the mass is

$$m = \rho \cdot (L)^3$$

where L is the unit length corresponding to either the prototype L_p or the model L_m. To have geometric similarity between the two systems the ratio of lengths must be constant everywhere; hence,

$$L_R = L_m / L_p \qquad \text{(A.2)}$$

This is called the model's scaling ratio.

Models corresponding to the prototype represented must be operated in *kinematic similarity*. This means that the ratio of the corresponding velocities in prototype and model must be constant throughout, that is,

$$V_R = \frac{V_m}{V_p} = \frac{L_m / T_m}{L_p / T_p} = \frac{L_R}{T_R}$$

where T_R is the time ratio between times in the prototype T_p and times in the model, T_m. To have kinematic similarity

$$T_R = T_m / T_p \qquad (A.3)$$

must be maintained.

Returning our attention to Equations A.1 and considering the succeeding formulas, we may conclude that to have *dynamic similarity* between model and prototype, the ratio between corresponding forces should be

$$F_R = \frac{F_m}{F_p} = \frac{m_m}{m_p} \cdot \frac{a_m}{a_p}$$

$$= \frac{\rho_m}{\rho_p} \cdot \left(\frac{L_m}{L_p} \right)^3 \cdot \frac{\Delta[L_R / T_R]}{\Delta T_R}$$

$$= \rho_R (L_R)^3 \frac{L_R}{T_R^2}$$

that is,

$$F_R = \rho_R (L_R)^2 (V_R)^2$$

in which ρ_R is the ratio of fluid densities in the prototype and model, and

$$(L_R)^2 = A_R$$

represents the ratio of the corresponding areas in model and prototype. The fundamental equation to assure dynamic similarity between prototype and model is, therefore the one representing the ratio of inertia forces,

$$F_R = \rho_R A_R V_R^2 \qquad (A.4)$$

In practical hydraulic structure design problems we are aware that the flow is influenced by a variety of forces. Inertia, gravity, viscous shear, capillarity, and elasticity all play a part. Usually the force of gravity or viscous shear forces are the most dominant. Because of the combined effect of the various forces acting in the prototype, it is practically impossible to design a hydraulic model that would be a true representative of the prototype for all forces present. It is therefore necessary to judiciously neglect forces of lesser importance and concentrate one's efforts on satisfying modeling requirements for the most important forces. For only one dominant force, neglecting all other forces, one should be able to build a model at a selected scale and operate it in such a manner that the effects of nondominant forces are negligible. In this way model studies can give reasonably accurate information on the expected hydraulic behavior of the prototype. If there are only two types of forces considered, Equation A.4 may be satisfied by selecting a model fluid different from water, which is the prototype fluid.

Most hydraulic models utilize water as a model fluid. Hence, the density ratio in Equation A.4 is unity. In some cases other fluids may be used to obtain better dynamic similarity. Once the density ratio is fixed Equation A.4 will be a function of two independent variables only: L_R and T_R. One provides for geometric similarity and the other for kinematic similarity.

Most hydraulic models represent either open channel flow, spillways, and weirs, for example, or pipe and closed conduit flow. The former are characterized by the work of gravity forces; the latter by viscous forces. In Chapter 8 the Froude number was introduced as a dimensionless parameter. It is the ratio of the inertia forces to the gravity forces. In Chapter 6 the Reynolds number was introduced to represent the dimensionless ratio of the inertia forces to the viscous forces.

To assure that the Froude numbers in model and prototype are the same, one may write the ratio of gravity forces as

$$\frac{\gamma_m}{\gamma_p} \cdot \left(\frac{L_m}{L_p}\right)^3 = \gamma_R (L_R)^3 \tag{A.5}$$

Since the Froude number is the ratio of gravity forces to inertia forces, the dynamic similarity formula, Equation A.4, should be equated with Equation A.5 because their ratio must be unity. This results in

$$\gamma_R (L_R)^3 = \frac{\gamma_R}{g_R} (L_r)^3 \frac{L_R}{T_R^2}$$

from where it follows that

$$T_R = \sqrt{L_R / g_R}$$

if the Froude numbers in model and prototype are to be similar. Since the model is under the same gravitational acceleration as the prototype, g_R is always unity. Hence for models of open channels and free surface flow structures the time ratio will take the form

$$T_R = \sqrt{L_R} \tag{A.6}$$

and the velocity ratio

$$V_R = \sqrt{L_R}$$

These equations represent the Froude model law for time scale.

When the flows are dominated by viscous shear, the model should operate such that the Reynolds number in the model corresponds to that in the prototype. Writing the ratio of viscous forces in model and prototype as

$$\mu_R \, (L_R)^2 \, / \, T_R$$

by Newton's law we may equate this with Equation A.4 since their ratio must be unity, that is,

$$\rho_R \frac{(L_R)^4}{(T_R)^2} = \mu_R \frac{(L_R)^2}{(T_R)}$$

From this we may express T_R after simplification as

$$T_R = (L_R)^2 \; \frac{\rho_R}{\mu_R} \tag{A.7}$$

which is the Reynolds model law for time scale.

From the two main model laws expressed by Equations A.6 and A.7 all other hydraulic parameters of dynamically similar models can be derived. For example, let us express the ratio of corresponding accelerations in model and prototype designed to satisfy the Reynolds model law:

$$a_R = \frac{a_{\text{model}}}{a_{\text{prototype}}} = \frac{L_R}{(T_R)^2}$$

By Equation A.7, T_R is substituted, and

$$a_R = \frac{L_R}{[(L_R)^2 \, \rho_R^2 \, / \, \mu_R]} = \frac{L_R \mu_R^2}{(L_R)^4 \rho_R^2}$$

Hence any acceleration measured in the prototype corresponds to model accelerations as

$$a_{\text{model}} = \left[\frac{\mu_R^2}{\rho_R^2 (L_R)^3} \right] \cdot a_{\text{prototype}} = \frac{\nu_R^2}{(L_R)^3} \cdot a_{\text{prototype}}$$

This formula may be referred to as the transfer formula between model and prototype. For the two main model laws these transfer formulas were determined for a number of hydraulic parameters that may be measured in prototype and model. The list of transfer formulas is shown in Table A2.1.

Table A2.1
Transfer Formulas for Froude and Reynolds Modeling Laws

Quantity	Modeling ratios for:	
	Froude law	Reynolds law
Length	L	L
Area	L^2	L^2
Volume	L^3	L^3
Mass	$L^3 \gamma g^{-1}$	$L^3 \gamma g^{-1}$
Density	γg^{-1}	γg^{-1}
Time	$L^{0.5} g^{-0.5}$	$L^2 \nu^{-1}$
Velocity	$L^{0.5} g^{0.5}$	$L^{-1} \nu$
Acceleration	g	$L^{-3} \nu^2$
Angular velocity	$l^{-0.5} g^{0.5}$	$L^{-2} \nu$
Angular acceleration	$L^{-1} g$	$L^{-4} \nu^2$
Force and weight	$L^3 \gamma$	$\nu^2 \gamma g^{-1}$
Pressure	$L \gamma$	$L^{-2} \nu^2 \gamma g^{-1}$
Impulse and momentum	$L^{3.5} \gamma g^{-0.5}$	$L^2 \nu \gamma g^{-1}$
Discharge, volume/sec	$L^{2.5} g^{0.5}$	$L \nu$
Discharge, weight/sec	$L^{2.5} \gamma g^{0.5}$	$L^{-2} \nu^3 \gamma g^{-1}$
Energy and work	$L^4 \gamma$	$L \nu^2 \gamma g^{-1}$
Power	$L^{3.5} \gamma g^{0.5}$	$L^{-1} \nu^3 \gamma g^{-1}$
Torque	$L^4 \gamma$	$L \nu^2 \gamma g^{-1}$
Absolute viscosity	$L^{1.5} \gamma g^{-0.5}$	$\nu \gamma g^{-1}$
Kinematic viscosity	$L^{1.5} g^{0.5}$	ν
Surface tension	$L^2 \gamma$	$L^{-1} \nu^2 \gamma g^{-1}$

The roughness of a model built either under the Froude or the Reynolds model law presents another design problem. The scaling ratio of the n roughness coefficient between model and prototype is determined by

$$n_R = (g_R)^{-1/2}(L_R)^{1/6}$$

For most models in which roughness may be a controlling factor the proper model roughness can be obtained by a trial and error procedure that should be carried out before the actual experimentation on the model begins. This involves observations of prototype behavior and adjustment of model roughness until the time ratio of the model and prototype corresponds to its desired value. River and harbor models are the most sensitive to these effects; hence their geometric distortion is usually controlled by their relative roughness.

Bibliography

Addison, H. *A Treatise on Applied Hydraulics.* (Fifth Ed.) London: Chapman & Hall Ltd., 1964.

Albertson, M. L., J. R. Barton, and **D. B. Simmons.** *Fluid Mechanics for Engineers.* Englewood Cliffs, N.J.: Prentice-Hall, Inc., 1960.

American Concrete Pipe Association. *Concrete Pipe Design Manual.* (First Ed.) 1970.

American Society of Civil Engineers. *Hydraulic Models,* ASCE Manual of Engineering Practice No. 25. New York, 1963.

American Society of Civil Engineers. *Hydrology Handbook,* ASCE Manual of Engineering Practice No. 28. New York, 1963.

Bean, H. S., ed. *Fluid Meters.* ASME Report. (Sixth Ed.) New York: American Society of Mechanical Engineers, 1971.

Cedergren, H. R. *Seepage, Drainage and Flow Nets.* New York: John Wiley & Sons, 1967.

Chow, V. T. *Open Channel Hydraulics.* New York: McGraw-Hill Book Co., 1959.

Davis, C. V. *Handbook of Applied Hydraulics.* (Second Ed.) New York: McGraw-Hill Book Co., 1952.

C. T. B. Donkin. *Elementary Practical Hydraulics of Flow in Pipes.* New York: Oxford University Press, 1959.

Fair, G. M., and **J. C. Geyer.** *Water Supply and Waste-Water Disposal.* New York: John Wiley & Sons, 1956.

Finch, V. C. *Pump Handbook.* California: National Press, 1948.

Fox, R. W., and **A. T. McDonald.** *Introduction to Fluid Mechanics.* New York: John Wiley & Sons, 1973.

Graf, W. H. *Hydraulics of Sediment Transport.* New York: McGraw-Hill Book Co., 1971.

Golzé, Alfred A. *Handbook of Dam Engineering.* New York: Van Nostrand Reinhold Co., 1977.

Harr, M. E. *Groundwater and Seepage.* New York: McGraw-Hill Book Co., 1962.

222

BIBLIOGRAPHY

Hicks, T. G., and **T. W. Edwards.** *Pump Application Engineering.* New York: McGraw-Hill Book Co., 1971.

Jaeger, Charles. *Engineering Fluid Mechanics.* Translated from the German by P. O. Wolf. London: Blackie & Son, Ltd., 1956.

Johnstone, D., and **W. P. Cross.** *Elements of Applied Hydrology.* New York: The Ronald Press Co., 1949.

Kaufmann, W. *Fluid Mechanics.* New York: McGraw-Hill Book Co., 1963.

King, H. W., and **E. F. Brater.** *Handbook of Hydraulics.* (Fifth Ed.) New York: McGraw-Hill Book Co., 1963.

Kinori, B. Z. *Manual of Surface Drainage Engineering,* Vol. 1. Amsterdam: Elsevier Publishing Co., 1970.

Lea, F. C. *Hydraulics for Engineers and Engineering Students.* (Fourth Ed.) London: Edward Arnold & Co., 1923.

Leliavsky, S. *River and Canal Hydraulics.* London: Chapman and Hall, Ltd., 1965.

Linsley, R. K., M. A. Kohler, and **J. L. H. Paulhus.** *Applied Hydrology.* New York: McGraw-Hill Book Co., 1949.

Linsley, R. K. and **J. B. Franzini.** *Water Resources Engineering.* (Third Ed.) New York: McGraw-Hill Book Co., 1979.

Luthin, J. N. *Drainage Engineering.* New York: John Wiley & Sons, Inc., 1966.

Medaugh, F. W. *Elementary Hydraulics.* New York: D. Van Nostrand Co., 1924.

Mosonyi, E. *Water Power Development,* Vols. 1 and 2. Budapest. Publishing House of Hungarian Academy of Sciences, 1963.

Morris, H. M. *Applied Hydraulics in Engineering.* New York: The Ronald Press Co., 1963.

O'Brien, M. P., and **George H. Hickox.** *Applied Fluid Mechanics.* New York: McGraw-Hill Book Co., 1937.

Owczarek, J. A. *Introduction to Fluid Mechanics.* Scranton, Pa.: International Textbook Co., 1968.

Parmakian, J. *Waterhammer Analysis.* Englewood Cliffs, N.J.: Prentice-Hall, Inc., 1955.

Portland Cement Association. *Handbook of Concrete Culvert Pipe Hydraulics.* 1964.

Powell, R. W. *An Elementary Text in Hydraulics and Fluid Mechanics.* New York: The Macmillan Co., 1951.

Powell, R. W. *Mechanics of Liquids.* New York: The Macmillan Co., 1940.

Raghunath, H. M. *Dimensional Analysis and Hydraulic Model Testing.* Asia Publishing House, 1967.

Rouse, H. *Engineering Hydraulics.* New York: John Wiley & Sons, 1953.

Russell, G. E. *Hydraulics.* (Fifth Ed.) New York: H. Holt and Co., 1942.

Sellin, R. H. J. *Flow in Channels.* London: Gordon and Breach Science Publishers, 1970.

Simon, A. L. *Practical Hydraulics.* (Second Ed.) New York, John Wiley and Sons, Inc. 1981.

Spink, L. K. *Principles and Practice of Flow Engineering.* (Ninth Ed.) The Foxboro Co., 1967.

Streeter, V. L. *Fluid Mechanics.* (Fourth Ed.) New York: McGraw-Hill Book Co., 1966.

Streeter, V. L. *Handbook of Fluid Dynamics.* New York: McGraw-Hill Book Co., 1961.

U.S. Department of Agriculture. *National Engineering Handbook.* Soil Conservation Service: Washington, D.C., 1971.

Vennard, J. K. *Elementary Fluid Mechanics.* (Third Ed.) New York, London: John Wiley & Sons; Chapman and Hall, Ltd., 1954.

Veeruijt, A. *Theory of Groundwater Flow.* New York: The Macmillan Co., 1970.

Woodward, S. M. and **C. J. Posey.** *Hydraulics of Steady Flow in Open Channels.* New York: John Wiley & Sons, 1941.

Index